SCIENTIFIC & MATHEMATICAL BODIES

NEW DIRECTIONS IN MATHEMATICS AND SCIENCE EDUCATION
Volume 22

Scope

Mathematics and science education are in a state of change. Received models of teaching, curriculum, and researching in the two fields are adopting and developing new ways of thinking about how people of all ages know, learn, and develop. The recent literature in both fields includes contributions focusing on issues and using theoretical frames that were unthinkable a decade ago. For example, we see an increase in the use of conceptual and methodological tools from anthropology and semiotics to understand how different forms of knowledge are interconnected, how students learn, how textbooks are written, etcetera. Science and mathematics educators also have turned to issues such as identity and emotion as salient to the way in which people of all ages display and develop knowledge and skills. And they use dialectical or phenomenological approaches to answer ever arising questions about learning and development in science and mathematics.

The purpose of this series is to encourage the publication of books that are close to the cutting edge of both fields. The series aims at becoming a leader in providing refreshing and bold new work—rather than out-of-date reproductions of past states of the art—shaping both fields more than reproducing them, thereby closing the traditional gap that exists between journal articles and books in terms of their salience about what is new. The series is intended not only to foster books concerned with knowing, learning, and teaching in school but also with doing and learning mathematics and science across the whole lifespan (e.g., science in kindergarten; mathematics at work); and it is to be a vehicle for publishing books that fall between the two domains—such as when scientists learn about graphs and graphing as part of their work.

Scientific & Mathematical Bodies

The Interface of Culture and Mind

By

SungWon Hwang
Nanyang Technological University, Singapore

and

Wolff-Michael Roth
University of Victoria, Canada

SENSE PUBLISHERS
ROTTERDAM / BOSTON / TAIPEI

A C.I.P. record for this book is available from the Library of Congress.

ISBN 978-94-6091-565-9 (paperback)
ISBN 978-94-6091-566-6 (hardback)
ISBN 978-94-6091-567-3 (e-book)

Published by: Sense Publishers,
P.O. Box 21858, 3001 AW Rotterdam, The Netherlands
www.sensepublishers.com

Printed on acid-free paper

CONTENTS

PREFACE

> A stone is the exteriority of singularity in what would have to be called its mineral or mechanical actuality *(litteralité)*. But I would no longer be a "human" if I did not have this exteriority "in me," in the form of the quasi-minerality of bone: I would no longer be a human if I *were* not a body, a spacing of all other bodies and a spacing of "me" in "me." A singularity is always a body, and all bodies are singularities (the bodies, their states, their movements, their transformations). (Nancy, 2000, p. 18)

In this introductory quote, Jean-Luc Nancy, perhaps the most prominent living French philosopher, points to the role of the body not only in knowing but also in being human tout court. It is the living body that makes me human, but within this living body, there is also an exteriority, such as a stone; and without this exteriority within me, I would not be able to relate to anything that is outside of my body, like the world. This living body in science and mathematics learning is precisely the central phenomenon of the present book. Whereas other science and mathematics educators focus on the mind and mental constructions or treat bodily experiences as stepping-stones to formal and abstract reasoning, we insist that without the sensuous, living and lived body, human beings would not be able to know anything at all. This book, therefore, is about the living and lived body in mind and culture with a particular perspective on science and mathematics.

Since the beginning of 2003, the two authors have worked together on a variety of projects, held together by a singular concern for the role of the body in communication, learning, and literacy. Our work has always been characterized by close attention to and work with data, which served for us as the ultimate testing ground for any theory that was to emerge from our work. We spent many hours, days, weeks, and months watching and analyzing videotapes featuring children and students of various ages engage with the tasks that their teachers (professors) had provided them with for the purpose of learning one or the other science or mathematics. This book constitutes the most current common conceptualizations that we have worked out over the past two years. In fact, we took the opportunity to take already published work and opened up our discussions for revising and reworking what we had done to take into account what we have learned since their original conception and publication. We found ourselves with words that we no longer are using today because of their problematic nature – e.g., "meaning" or "construction" – and we replaced them as part of our conceptualization of this book.

In this book, we draw on the original versions of French and German philosophical texts. When we quote from these texts, all translations are our own, though, where available, we have checked our version with the one published in the English language. Because of the non-equivalence of languages, translators have to make choices, which generally orient themselves to the intended audiences – philosophers. In our translations we have attended to the suitability of the translation for our audience: mathematics and science educators.

The original texts underlying the eight chapters of this book have previously been published in conference proceedings and academic journals before we further and extensively revised and expanded them for this book project. The first versions of chapters 1 and 3 have been presented at the first and second *International Science Education Conference*, which were held in 2006 and 2009 at the National Institute of Education, Singapore. The annual meetings of the *American Educational Research Association* in 2008 and 2010 provided the opportunities to present and discuss the initial versions of chapters 2 and 5 of which the former was subsequently published in *Research in Science Education.* An article that appeared in *Pedagogies: An International Journal* served as the basis for chapter 4. *Cultural Studies of Science Education* published ideas discussed in chapter 6 and 7 and earlier on, the latter had been presented as part of the first *Springer Forum on Science Education* on the topic of globalization (Chicago, 2007). Chapter 8 first appeared in *Forum Qualitative Sozialforschung / Forum Qualitative Social Research.*

This book would not have been possible without various research grants and supports from individuals. Several research grants from the Social Sciences and Humanities Research Council of Canada (to W.-M. Roth) enabled collecting the data for all chapters and supported the writing of the initial papers and presentations. We thank our research participants for making their everyday lives available to us and welcoming our presence in their settings. A research grant from the Office of Educational Research, National Institute of Education, Singapore (OER 7/09 HS) provided the opportunity to plan this book project and develop the integrated framework for encompassing the entire chapters. We are grateful to the individuals who contributed to the earlier drafts of the chapters. Mijung Kim co-authored the initial versions of chapters 1, 4, 5, and 6 and collaborated on the OER research project. The members of CHAT@UVic provided SungWon Hwang with numerous discussion opportunities during her stay in Victoria in 2007. These individuals include Peilan Chen, Michiel W. van Eijck, Gholamreza Emad, Maria Inês Mafra Goulart, Pei-Ling Hsu, Bruno Jayme, Jean-François Maheux, Giuliano Reis, and Eduardo Sarquis Soares.

Singapore and Victoria
January 2011

TOWARD A THEORY OF THE BODY IN SCIENTIFIC AND MATHEMATICAL COGNITION

> And how will you enquire, Socrates, into that which you do not know? What will you put forth as the subject of enquiry? And if you find what you want, how will you ever know that this is the thing which you did not know? (Plato, 380 BCE)

Reason, the human capacity to make sense of the world, has long been the goal of science and mathematics education. Science and mathematics educators have taken forms of reasoning – including questioning, hypothesizing, and justifying – with empirical data central to cultural development of individuals and societies. Investigations of how people know and learn have been conducted from a variety of theoretical perspectives. Traditionally the discussion of reason often has been restricted to accounts constructed after the sense-making event as if a thought were a thing that could stand independently of the concrete praxis of talking, writing, or doing in a particular setting. More recently, the notion of reason has been expanded to that which is inseparably conditioned by culture and language and also the body engaged in the praxis. The cultural nature of knowing science and mathematics (i.e., social constructivism) and the primary role of the body in shaping the mind (i.e., embodiment) contributed to incorporating the *actual* ways in which humans act and interact in and with the everyday world. Crucial to this transition from the static (absolute) notion of reason to the *living* process of sense making is the acknowledgement of the mundanity of knowing and learning science and mathematics, which pertains to the ontological conditions that also have made possible the establishment and development of disciplines such as science and mathematics. The ordinary, taken-for-granted, everyday life that human agency is inescapably caught up in materially and culturally while dong and talking science or mathematics has been acknowledged as an inseparable and central moment of scientific and mathematical sense making.

The theoretical orientation to everydayness in learning theories commonly admits that the human being is not a machine, like a computer, which mechanically transfers information. Rather, a recognition now emerges that being human means being in the flesh, acting in the world with feelings, emotions, and corporeal forms of knowing. The classical theory of reason, which explains learning as the product of the conscious (rational) mind configuring itself, does not fully explain how real people learn by means of sense experiences, affect, and uncertainties. Like the paradox that Meno articulates for his mentor Socrates in the opening quote of this introduction, knowing and learning theorized in terms of a rational mind would come to an unsolvable problem: We cannot take the unknown thing as the object of an intentional learning act: How can we aim at learning something that we do not even know that it exists? If learning consists of rational processes despite this inescapable problem, this learning paradox, then it is because humans have a world of possibilities ready at hand in its mundanity. The unknown learning object reveals and gives itself as learners engage in activity the purpose of which is not entirely clear to them. For example, students learning Newtonian mechanics for the first time in their lives do not directly encounter "abstract" physics concepts (e.g., inertia) but come face to face with concrete human artifacts including mathematical equations, visual representations, scientific equipment, and text- or sound-words. For students who do not yet grasp a concept, who do not even know what the intended concept involves, there is no magical solution for learning scientific concepts such as inertia. They can only engage in talking with and about these concrete objects (e.g., seeing, hearing, and touching) and in letting sense appear to them from this everyday (ordinary) experience of the world. Thus, students doing an experiment in a science laboratory grapple with mathematical equations and scientific equipment, and thereby evolve a better sense of some scientific phenomenon. In this way, learning science and mathematics occurs in their everyday lifeworlds rather than in some metaphysical conceptual netherworld abstracted from reality. Students' power to act knowledgeably in their familiar world is inseparably intertwined with their everyday experiences. Everydayness, which refers to the condition that real people (embodied creatures) inhabit in and for their everyday practice, constitutes both the context of and resource for expanding the sense of the world and therefore for learning science and mathematics.

MIND IN CULTURE = CULTURE IN MIND

> [A]ll higher functions evolve in phylogeny not biologically, but social; (4) the crudest meaning – the mechanism of such functions is a copy of the social. *They are internalized relations of a social order, transferred to the individual personality,* the basis of the social structure of the personality. Their composition, genesis, and function (mode of action) – in a word, their *nature – are* social. Even transformed in the personality into psychological processes, they remain quasi-social. (Vygotsky, 1989, pp. 58–59)

Contemporary sociocultural theories of science and mathematics education use the term *literacy* to conceptualize the goal of teaching and learning science and mathematics (i.e., scientific or mathematical literacy). As an extension of the initial sense of the term "knowledge of letters" and "being able to read and write," the notion of literacy generally accepted in education research, practice, and policymaking refers to a cultural competency. This competency is not limited to the traditional notion of knowledge (e.g., pieces of information) and (transcendental) rationality but indicates the competency of communication and thinking in and across various cultural settings (e.g., school, workplace, or place of residence). Rather than the simple acquisition of information, the notion of literacy acknowledges that learning science or mathematics means becoming part of a community of cultural (linguistic) practice. That is, learning science and mathematics is like learning a new (foreign) language. This learning is not unlike what Lev Vygotsky (1934/1986) proposed a long time ago in his studies of children's scientific concept development. This form of growth requires the development of everyday rationality, which means that one participates in talking science and mathematics and becomes knowledgeable in her ordinary, day-to-day life. The equivalence of learning science or mathematics to learning a new language – implied also by the linguistic root of the term *literacy* – does not only mean that educators need to attend to everyday language and linguistic resources deployed in the communication of new scientific concepts. More than that, the core issue of linguistic and cultural approach to learning science is that learning a new (foreign) language presupposes the human capacity to know the world and going beyond the learner's current understanding. Learning a language is not a purely linguistic process. For example, anyone who has had the opportunity to observe a child learning a language sees that a newborn comes to be able to speak a language not only because someone tells her how to speak or what the sense of each word is but also because a child has opportunities to be part of everyday events that are brought about in and through the use of the language. Language and life are interwoven like warp and weft, and pulling one means that the cloth falls apart without anything left. From a pragmatic perspective, there really is no difference between knowing a language and knowing one's way around the world generally. The difference between the two – knowing a language and knowing one's way around the world – is undecidable.

Anyone having experienced the learning of a language other than her mother tongue knows that one does not grasp the sense of a foreign word through reading the word definition in a dictionary until she experiences the word really in making an event happen or getting things done. For example, one of our research projects took us into a fish hatchery. Initially, everything appeared foreign and strange and we did not understand what the inhabitants of this place were talking about. But over time, we not only knew what they were saying but we were also able to replace them in their daily tasks. Perhaps this is why students may mechanically memorize the textbook definition of scientific or mathematical terms but this does not necessarily show that they can cope with real, everyday events.

To sum up, we propose thinking and theorizing scientific and mathematical sense making in terms of two ideas: learning science and mathematics is comparable to learning a foreign language, on the one hand, and knowing a language is equivalent to knowing one's way of living in the world, on the other hand. The following two propositions summarize the two moments of sense making from a cultural (literacy) approach:

a. Knowing science and mathematics means knowing ways of talking science and mathematics and knowing to talk about scientific and mathematical objects.

b. Knowing a language is equivalent to knowing one's way around the world. That is, knowing how to talk scientifically or mathematically means knowing one's world around science or mathematics.

The two statements, each of which describes a different aspect of knowing (language, practical sense of life), are not independent but interrelated. Once the role of language (communication) is taken for granted (as a transparent medium between statement a and b) we are led to notice that the cultural significance of learning science and mathematics lies in expanding one's room to maneuver in the world. At the same time, one expands one's ways of getting around the world (i.e., collective social life) by learning to knowledgeably participate in talking science or mathematics. Many scholars in the fields of language philosophy, phenomenological sociology, and cognitive psychology point out the mutually constitutive relation between knowing how to speak a language and knowing one's way around the world. Thus, an individual whose major is social science may experience physics as a strange culture and as a very different language to learn. She may decide to study physics because of her enthusiasm about knowing the beauty of physics. However, to be knowledgeable about the field she needs to participate in various practices and find her way of getting around this culture. This is what we could observe when we followed one student for an extended period of time across very different activities including lectures, doing homework, consulting with a lecturer, doing a group project, presenting a project in class, and doing laboratory work (Figure 0.1). She developed familiarity with physics as she engaged in the activities that played themselves out in the different places that constitute a university physics department.

BODY IN MIND = MIND IN BODY

> There is an experience of the visible thing as pre-existing my vision, but it is not fusion, coincidence: because my eyes that see, my hands that touch, can also be seen and touched, because, therefore, in this sense, they see and touch the visible, the tangible, from the inside, that our flesh pervades and even envelops all things visible and tangible of which it is nevertheless surrounded, the world and I are one in the other, and of the *percipere* to the *percipi* there is not anteriority, there is simultaneity to the same delay. (Merleau-Ponty, 1964, p. 162)
>
> The flesh [is the] immemorial memory of the world. (Henry, 2000, p. 206)

Both opening quotes lead us to the core issue of this book: the living body, flesh, as the necessary condition for knowing, memory, and representation. Without the living body, there is no consciousness of life to become conscious of itself. Flesh *is* the immemorial memory so that body and mind are two manifestations of the same thing: the flesh. The two moments of knowing science or mathematics have been topics of phenomenological studies over the past decades. These studies – unlike science or mathematics education research that is concerned with stuff in the heads of students – are interested in the way in which students are situated in their life-

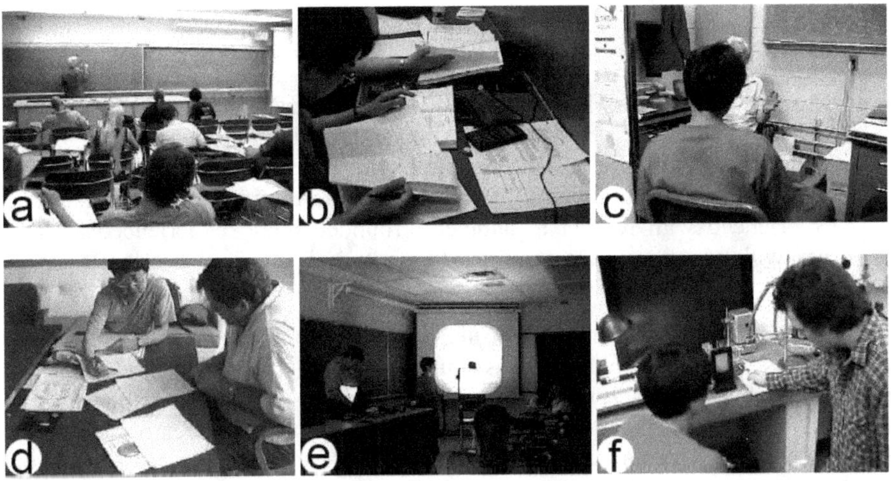

Figure 0.1. A third-year undergraduate physics student participates in various everyday events of learning physics. Every event is driven by internal contraction between two moments of learning in everydayness (communication, practical sense development). The bodies that act, which are observable not only to the members of the collective practice but also to analysts (audience), mediate communication, experience of the world, and the dynamic relation between the two. a. Attending a lecture. b. Doing homework with a peer. c. Consulting with a lecturer to get additional assistance. d. Conducting a group project. e. Presenting a group project in a class. f. Doing a laboratory work.

worlds. This research is interested in learning from the perspective of the individual student, in the way that the world appears to her, her needs, and so on (e.g., Roth, McRobbie, Lucas, & Boutonné, 1997).

A few educational researchers have acknowledged the role of the human body in learning science and mathematics. Studies on gestures and on the multimodality of communication have shown the crucial role of the body in teaching and learning. Of these, an increasing number of investigations have taken an anthropological framework with respect to the question of what it means to be part of a cultural practice and to be knowledgeable about science or mathematics. These studies have theorized the bodily dimension of human practice and, therefore, of human knowing which researchers also approach from the perspective of the embodied mind (e.g., Lakoff & Núñez, 2000). Whereas traditional psychology conceives of knowing (and learning) as a mental phenomenon (e.g., information processing) to which the body is subordinate or constitutes only an external condition, some contemporary theories accept that the body is a constitutive moment of thinking. What and how we know involves the body. This means that once one pulls bodily knowing from the equation, there would not be any concepts left (e.g., Roth & Thom, 2009a). Therefore, the body is essential to conceptual communication, on the one hand, and to the experience of the world, on the other hand. The following two statements propose that the body is an irreducible moment of what we know as knowing – even though we might fail to acknowledge it. The two statements are:

a. The body constitutes hub relating inside and outside in producing communication, that is, in hearing, speaking, and making sense.
b. The body constitutes the hub relating inside and outside in the experience of the world and its objects: "Relative is a movement necessarily that is experienced with respect to another 'grounded body' experienced as in rest, with which my own corporeal living body is one." (Husserl, 1940, p. 311)

The two propositions complement the previous two about knowing science or mathematics, which exhibits itself as mundane knowledgeability in cultural and linguistic events. Together, the two sets of propositions provide an explanation of how individual minds make contact with culture. Human beings corporeally (e.g., visually) communicate and experience the world before they are capable of talking about it. If they were talking about a world they do not know, they literally would not know what they are talking about. For example, in/with my sensuous, living and lived body that is capable of knowing as much as of suffering I take up a spatiotemporal position at every instance of my life. (We follow the general approach in the phenomenological literature to use the first person pronoun, because to the individual, the body of reference is always mine. It is with respect to my body that I experience the spatiality and temporality of the world [Husserl, 1940].) This taking a position constitutes the condition for my everyday knowing, experience of time and space, and communication (i.e., the body as the very condition for taking a standpoint). My position not only determines my disposition, but also makes me subject to exposure. Through my body, as Merleau-Ponty points out in the introductory quote to this section, I perceive the material world in a specific way of which the (perception) process also structures my body and allows my body to sense the world differently. It is precisely this subjectivity that I share with others as well as the knowing about this subjectivity (rather than the contents of our knowing). This explains students' concept development, for example, pertaining to the question of how they come to see (hear, feel) something that has not been seen (heard, felt) before. The expression I produce allows me to see/hear myself thinking, and this, according to Vygotsky (1934/1986), allows me to change my thinking. One might therefore ask how a third-year university student develops when she writes up an experiment she has done in a laboratory report, and how she is learning should she recognize her graph (Figure 0.2) as standing in for and exhibiting the cyclic motion of an engine. But she – as any other human being would be – is unable to anticipate this future knowing, because such anticipation requires knowledge of the knowledge. Or, to state the problematic in the terms of more classical theories, anticipation requires *re*presentations and these *re*presentations precisely constitute what does not yet exist for the learner, who, again in traditional terms, "first has to construct these representations."

The phenomenological approach to culture opens up new avenues for thinking our sensuous bodies as the places where culture and mind are articulated. My body

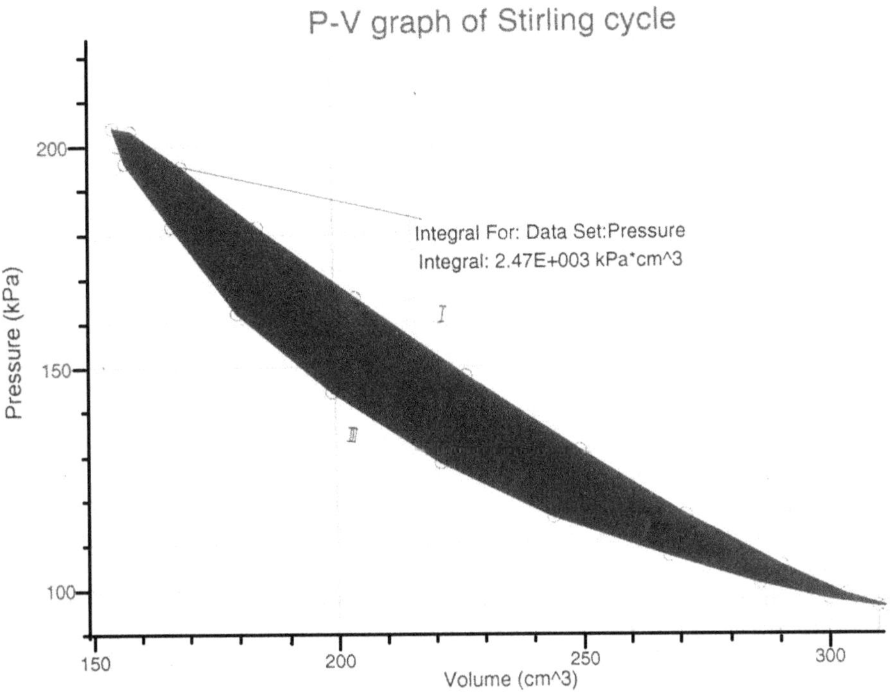

Figure 0.2. An excerpt from a physics laboratory report written by a third-grade undergraduate physic student: If she sees in this graph a cyclic motion of a thermodynamic engine, what would make it possible?

is open to the social and material world, and, because of its senses, it is subject to be affected (Bourdieu, 1997). My body constitutes the interfaces between culture and my mind, which mediates the everydayness of communication and the mundanity of experiencing the world in various situations that I come to be in when I learn science or mathematics. In fact, there is no difference between mind and culture (society), for culture is in the mind as much as mind is in culture. If it were not like this, no cultural learning would be possible and nobody else would be able to recognize in my doing something that they also could do. The general (virtual) possibilities of culture exist concretely only in the real, corporeal action of individuals such as myself.

My body is constitutive of my participation in everyday practice. It has been noted that all higher psychological functions have been social relations first (Vygotsky, 1989). While taking part in science or mathematics learning tasks, students engage in relations so that what they know as scientific and mathematical ideas simply is a reflection of these relations. The mediating role of my body as the interface with others and the world means that it is actively involved in inscribing the sense of things (bodies) including language, mathematical representations, and scientific equipment. My body therefore is integral to my sense making. Following phenomenological philosophers my living body – that is, my body that is capable of suffering, of being affected, of affect – is the source of sense. In the phenomenological literature, this body is referred to as "the flesh" (e.g., Merleau-Ponty, 1945). The living body relates to other scientific or mathematical bodies and simultaneously translates some communicative actions into forms of knowing the world and some of forms of experiencing the world into communicative actions.

The incorporation of the body into a theoretical framework of sense making is consistent with a cultural-historical psychology that aims to develop a comprehensive theory of thought and communication (Vygotsky, 1934/1986). Vygotsky points out the existence of a dynamic relation between communication and conception by theorizing the phenomenon of "word-meaning" and its development. Because the term "meaning" derives from a primitive conception of language, however, we do not use this term in this book (unless in quotation marks to refer to the use of the term by other scholars). Rather, instead of Vygotsky's term "word-meaning," we draw on the equivalent concept of *participative (unindifferent) understanding* (Bakhtin, 1993). Instead of "meaning" we sometimes also use "signification" or "sense," terms that are consistent with the French and German translations of the Bakhtin's work and with its origin in the writings of Ferdinand de Saussure.

The dynamic relation of participative (unindifferent) understanding further develops to the concept of growth point in psycholinguistics (McNeill, 2002). Grounded in a dialectical theory, the growth point of an idea is analogous to the seed of a tree. It is only a beginning in which the mature thing cannot yet be seen. Moreover, what this mature thing will look like depends on contextual factors that influence the phenotype of the growing idea (tree). The following two propositions summarize the central role of the living body in the dynamic process of sense making at the interface of culture and mind:

a. Communication is distributed within a communication whole that is produced by the sensuous, living/lived body (e.g., words, gestures, body movements, and prosody).
b. Gestures, body orientations, positions, and movement, prosody, and other communicative modalities constitute one-sided manifestations of an irreducible communicative whole.

First, the sense of communication is made available in and through actions such as sound-words, gestures, eye gaze, body positions, and body movements. That is, any one of them is not additional or subordinate to the other moments. Communication is distributed over the entire communicative unit that encompasses them. In fact, from a phenomenological perspective, my body in its entirety is the expression of thought; there is no thought apart from what my body expresses for others and for myself.

Second, these performances are observable not exclusively in communication. They are also constitutive of the ways by which people encounter and undergo during events in which science or mathematics is context or topic. I experience the world with my words, hands, eyes, and body. This is also the way in which any other "I" expresses itself. This double function of my living body implies that an expression made toward the other is potentially the way by means of which I practically experience the world. In fact, both Merleau-Ponty and Vygotsky point out that it is through our expressions that I come to know my own thoughts. Initially, when I begin to express myself, these thoughts consist only in nuclear and undeveloped form. As it were, I discover what I think precisely only in speaking. That is, my bodily expression *precedes* the recognition of the thought said to lie behind it. Every thought, in its mature form, literally is an afterthought, in its seed form at the moment of the growth point.

BODY | MIND | CULTURE

> Certainly, there is the question of knowing how the "ideas of intelligence" install themselves from above, how we get from the ideality of the horizon to "pure" ideality, and by means of which to the miracle notably a created generality adds itself to the generality of my body and the world, a culture, a knowledge that takes up and rectifies the first. (Merleau-Ponty, 1964, p. 197)

In this introductory quote, Merleau-Ponty problematizes a question that many critics of the embodiment approach legitimately ask. How do you get from the experience of the body, which is always a singular experience, to the generalities of the culture? Many years later, using almost the same language, Bourdieu (1997) would articulate the living body as the condition for culture. Impressible because of its senses, my living body is fashioned by and according to culture and material world such that a homology develops between the experienced and experiencing structures.

When we talk about knowing, developing, and learning then we presuppose body, mind, and culture. There is no learning or development without the living bodies, which are endowed with senses and therefore with the resources for making sense. There is no learning and development without the mind, which in fact is a feature that provided an evolutionary advantage and that natural selection therefore chose. There is no learning and development without culture, because it is only in collectivity that we have *con*sciousness (= *con-*, with + *sciere*, to know), a language, a means of handing down knowledge and practices to future generation – including the use of tools, such as physical apparatus, geometrical objects, or language.

The practice with physical artifacts may also constitute an expression toward the other. This is so because the (same) living body expresses, and thereby translates experiences (or acts and translates expressions). Therefore, making thematic the role of my sensuous, living and lived body in my sense making has the potential to lead to a comprehensive perspective on everyday learning in science or mathematics. Such a perspective does not dichotomize the different constituents in a phenomenon, such as cognition and emotion, everyday and scientific concepts, and individual and collective subjects. Thus, a theory of the living body constitutes a theoretical alternative to constructivism and other psychological orientations, which conceive of the rational mind as the only source of sense. The living human being, who experiences, emotes, feels, and develops, constitutes a firm theoretical or practical basis for a theory of rationality that is grounded in the mundane ways in which I actually experience myself. The following three assertions summarize the advantage of taking the living body as the center of a theory of sense making: radical passivity, heterogeneity, and solidarity.

> *Assertion 1 As we participate in the everyday world, we cannot anticipate what we do not yet know, including what we will be seeing and the sense we will be making. That is, we are in a situation of radical passivity with respect to absolutely new ways of perceiving and with the sense we can make of it.*

The shift made from the transmission model of learning to students' agency to construct cultural knowledge (in social interaction) has been the major achievement of science and mathematics education theory over the past several decades. Yet, the relation between culture and the individual mind, for example, how individual students *discover* the knowledge that humans have developed over the long cultural-historical time span has been one of the critical questions posed to constructivists long before this movement came about in science and mathematics education – as our introductory quote shows. In quite a number of studies, learning is considered as if it were a process of constructing a building with a set of blocks and students are assumed to see these blocks without any problem. That is, the child is said to know what kind of blocks there are in her toy box and, if needed, is able to display and see them all at once as competently as curriculum designers. If this is the real case of sense making, learning would simply be consisting of selecting some blocks among the whole set and constructing a building according to a plan. However, this metaphor has limitations (Roth, 2011b): When students learn about something really new, they actually cannot see either what building they are going to construct or what blocks are available for them yet. Both the building and the block pieces are invisible.

For this problem the theory of the living body provides a solution: sense making is possible because the child is capable of feeling (suffering) something that happens to or comes into contact with the body – being touched. Students actively engage in knowing the world by making their living bodies contact with the world while seeing, hearing, touching, and moving. Becoming aware of something through the senses (i.e., perception) is a by-product given to students in the course of their sensuous labor. That is, sense making involves a radically passive moment in which acting constitutes the possibility to be affected by the unknown and therefore not-yet-known, by the unseen and therefore unforeseen. This experience is similar to learning a new language and therefore to make a completely new sounds. In fact, a new language speaks through the person one day when she finds herself being able to pronounce the sound that she could not make before. This is also similar to traveling in an area that is totally new to a person. The new world (objects) appears to the person's vision one day when she finds herself seeing things that have not been visible before.

Assertion 2 We communicate by various modalities including sound (words), intonation, hand-arm movements (gestures), body movements, body positions. These modalities communicate differently (form) and are different (things). Communication therefore is heterogeneous. We cannot conflate all other modes to the verbal but need to seek the unity at a higher level. From the perspective of the higher level, all forms of expression are one-sided manifestations of the whole – and therefore constitute only partial truths.

The significant role of our bodies in sense making has been acknowledged in some educational studies in such key terms as *embodied knowing*, *situated learning*, *learning-in-practice*, and so on. Common to these concepts is the inseparable relation between individual knowing (e.g., being able to do arithmetic) and social and material context in which this knowing plays itself out (e.g., in grocery shopping). That is, neither the mind nor the context can be a complete unit for understanding everyday rationality other than the *praxis* that encompasses the dynamic relation between the two. Thus, communication is not a simple dumping of mental content into the public arena of a conversation. Rather, it is a process in which talking, gesturing, placing bodies, moving bodies, and intonating bring out contextual possibilities. Simultaneously, these contextual possibilities change, as talking adds new resources for thinking and for further talking.

This way of understanding provides a different perspective of communication: the role of communication in sense making is not to code-switch from one way of talking (e.g., the vernacular) to another (e.g., the scientific/mathematical) but to produce resources that problematize the boundary between the two. Communication brings about culture not in a self-identical manner but in heterogeneous ways. In the classroom, the heterogeneous nature of language and culture constitutes the very condition for knowing and learning science and mathematics. In fact, the double role of my living body in communication and practical experience of the world indicates two (inseparable) roots of sense making resources and therefore their heterogeneous nature. The way an individual participates in communication and learns should be studied by considering all the practical actions involving not only speaking/listening but also the interaction with visual and textual representations mobilized as part of the ongoing communication. The acknowledgement of the heterogeneous communicative performances as legitimate sense making resources brings about a different understanding of culture and language – language and culture are inherently heterogeneous because of the very possibility of their role in making sense of the world.

Assertion 3 Language, sense, consciousness, and knowledge inherently are cultural *(societal, collective) forms. They exist only in and as collective forms. Without the collectivity, there is no language, sense,* consciousness, *or knowledge. Collectivity, however, is equivalent to solidarity, which thereby is a condition of sense making.*

The double function of my body across two developmental lines – communication along the line of linguistic practice, on the one hand, and experience of the world along the line of object-oriented practice, on the other hand – raises a question: What condition makes this encounter possible on the first place? The *living* body that is distinguished from the material body by the fact of its being alive (i.e., life) constitutes the answer: The material body and the living body (flesh) are two moments of praxis for expanding one's condition of life. Making sense of the world does not mean practicing individual intelligence but it means improving the collective condition that also makes the individual intelligence possible in the first place. That is, the power to make sense of the world derives from the praxis that expands the collective condition of life. To live life means that I have to fulfill the basic needs that come with my body, food, shelter, clothing, love, and so on. Affect is a measure to which the basic and extended needs are fulfilled. All my actions in this world not only are reflected in (verbal) consciousness but also in affective terms. We can understand learning and development only when we take the perspective of the whole person, in her flesh and blood, and recognize that cognition is only a one-sided part of this whole (Vygotsky, 1989).

The perspective of the sensuous, living and lived body theorizes sense making in its unity with affect. We thereby respond to Vygotsky (1934/1986), who suggests that "when we approach the problem of interrelation between thought and language and other aspects of mind, the first question that arises is that of intellect and affect" (p.10). Life is realized in my sensuous body so that it is precisely in and through my sensuous body, its affectability, that I share life with others (Henry, 1990). Emotions are a direct reflection of our situation in the world – I do not have to think but experience fear in certain situations and I tend to flush before I am conscious of being embarrassed. Cultural-historical psychology finds the conditions for this inseparability in the orientation toward a collective motive that any participation inherently presupposes and realizes in and through a concrete action (Leont'ev, 1978). Collective emotions are always part of social interaction precisely because life is realized in a sensuous, affective body – my own emotions can change when I have a good time, when my team goes ahead in a game, or when my nation wins another gold medal. This approach opens possibilities to study teaching and learning in science and mathematics. Sense making changes the condition for knowing my way around the world and therefore changes the way in which I experience the world as well.

OVERVIEW OF THE CHAPTERS

This book consists of eight chapters grouped into three sections, each of which fleshes out the dialectic approach to sense making. Although science and mathematics educators will find the topics familiar, they will also note that we introduce new ways of thinking about them. In Part A, we articulate the double relation of body and sense: there is a sense of the body and a body of sense. The two moments of sense (ideal, material) are theorized to constitute one another – rather than independent – in and through our living bodies. In chapter 1 we articulate how the category of the living body realizes and extends existing approaches to participative (unindifferent) understanding and therefore succeeds in responding to the learning paradox – how learners can arrive at forms of knowledge of complexities higher than their ground, material, and tools. (The constructivist approach currently does not have an answer to the learning paradox, even though some eminent scholars including Ernst von Glasersfeld and Les Steffe have tried their hands at it.) The "construction of meaning" tends to be the main pedagogical goal for the teaching of concepts. Yet, the metaphor of "construction," which is also used to theorize the development of participative (unindifferent) understanding, articulates only some among the different possible forms of knowing. The purpose of this chapter is to articulate a more comprehensive approach to development than exists in Vygotsky's framework and its logocentric approach. That is, this study takes Vygotsky's theory and develops it to include not only the words that he focused on but also the living body as a whole to arrive at a more holistic approach. We draw on phenomenology to develop a new way of understanding the nature of concepts. We substantiate our theoretical position by presenting exemplary case studies that exhibit the children's trajectory of learning geometry in a second-grade mathematics classroom. We conclude that the extended framework of participative (unindifferent) understanding allows theorizing the role of emotion in learning mathematics without separating it from cognition.

In chapter 2, the living body is proposed as a category that captures lecturing activity better than information processing approaches – many science professors tell us that they try to "get" their "points across" – and therefore as the venue for conceptual gains. Lectures are often thought of in terms of information transfer: students (do not) "get" or "construct meaning of" what physics professors (lecturers) say and the notes they put on the chalkboard (overhead transparency). But this information transfer view does not explain, for example, why students appear to have a clear sense of understanding while they sit in a lecture and their subsequent experiences of failure to understand their own lecture notes or textbooks while preparing for an exam. Based on more than a decade of studies on the bodily nature of science lectures, the purpose of this chapter is to articulate and exemplify a different way of understanding physics lectures. We show that there is more to lectures than the talk plus notes. This "more" may indeed explain (part of) the gap between students' participative understanding that exists in the situation where they sit in the lecture, on the one hand, and the one where they study for an exam from their lecture notes, on the other hand. Our results suggest that in lectures, concepts are heterogeneous performances in which sense arises from the synergistic and irreducible transactions of many different communicative modes, including gestures, body movements, body positions, prosody, and so forth.

In Part B, the role of the body in sense making is articulated and elaborated. Sense making traditionally is studied only from the perspective of sense as a mental issue, when in fact, without the body, there would be no senses and therefore there would be no sense whatsoever (for example, computers cannot make sense but only transmit and process signals). Only a living body has senses and thereby has the means to make sense of the world. The four chapters in this section attend to the embodied and bodily nature of sense making at different settings of learning. Chapter 3 theorizes literacy as the performances of living bodies rather than being the outflow of mental or purely linguistic capacities. The body is deeply involved in knowing a language – my sensuous, living body is the link between speaking/hearing, on the one hand, and knowing my way around the world, on the other hand. Literacy denotes the cultural nature of knowing to communicate. Most studies of (scientific) literacy tend to use the notion of "constructing meaning" to theorize the process by means of which students become literate. Yet, the conditions that make literacy practice possible and bring about the associated conceptual development are hardly made salient and theorized. Therefore, literacy is often attributed to categories (e.g., mental) independent of bodily experience. In this chapter, we develop the idea that literacy always expresses itself somehow, concretely, in the (discursive) actions that are bodily produced and performed. We exemplify the central role of the body in computer-mediated literacy practice and the associated development of conceptual understanding in physics laboratory to support two claims about learning physics in a thermodynamics laboratory. First, the body temporally engages with different laboratory artifacts and spatially realizes literacy. Second, the body spatiotemporally coordinates cultural artifacts, which constitutes a terrain from which a higher-order cognitive function arises. We conclude that the central aspect of literacy in students' conceptual development consists in the performance of a world, which bridges interpretive resources and the experience of the world metonymically.

As part of chapter 4, we expand the relation between communication and the experience of the world and therefore theorize the role of living bodies in conceptual development. Existing pedagogical theories often posit the source of conceptual development in terms of the literacy of an individual child or teacher and therefore suggest instructional strategies or instructional models based on (frequently unquestioned) presuppositions about the learning process. These theories give little attention to the real conditions that make the (real-time) communication of an idea possible. In this chapter, we extend our approach to literacy and participative (unindifferent) understanding to give the body a central role in conceptual development. We exemplify the contentions that (a) the body constitutes the mediating hub that translates, in an ongoing manner, the experience of the world (objects) and the word in communication; and that (b) the body constitutes the mediating hub that generates a new ways of making sense of the word. We conclude that a body-centered framework helps understand the sensuous labor that increases the resources for translating participative (unindifferent) understanding.

In chapter 5, we focus on the learning paradox involved in knowing mathematical inscriptions (sign forms other than language used in [electronic, paper-and-pencil, or whiteboard] written communication) and study the role of the body. Science teachers frequently use mathematical inscriptions (representations) in their lessons. In classroom talk, different types of mathematical representations assist science teachers in explaining concepts. Yet, students often have difficulty reading (interpreting) mathematical inscriptions that they encounter in their science lessons and associated science conversations. In this study, we theorize the practice of reading mathematical representations from a holistic perspective and articulate pedagogical principles of concept development. We thematize the cultural dynamics of reading mathematical representation and exemplify it by analyzing a case example in which children in and through communication first come to know some geometrical inscription. First, the body reproduces and transforms cultural resources for interpreting mathematical representations. Second, the increase of heterogeneous interpretive resources in communication increases possibilities for realizing a new way of talking. Third, knowing mathematical inscriptions emerges from the different, irreducible modes of communication as an integrated whole. We conclude that knowing mathematical inscriptions is equivalent to (bodily and embodied) reading between elements of mathematical inscriptions that constitute a series of references to bring about scientific conceptions in its totality.

In the course of chapter 6, we theorize the role of the body in making sense of lecture talk. In this chapter, we suggest an approach to conceptual development that does not start from the dichotomy of the formal and the informal. We propose an approach that points out the problem of theorizing cognition (conceptual understanding) by depending on specific forms of representation (e.g., scientific terms). As alternative, we envision a cultural-historical approach to language, which considers different, irreducible modes of communication as an integrated whole and therefore allows theorizing cognition without separating it from the concrete ways by means of which human beings communicate. We provide an exemplary analysis of a lecture talk in a university physics classroom and exemplify dialectic theories that explain the development of conceptual understanding. We suggest that this approach takes account of the fact that people communicate scientific concepts through hybridization, which does not reproduce a genre self-identically but through the proliferation of difference.

In Part C, we provide perspectives on mathematical and scientific learning in terms of living bodies that inhabit the world where they encounter science or mathematics as new linguistic and cultural forms. The two chapters in this section provide an analytic framework of communication that makes emotional-volitional valence and ethico-moral valence integral aspects of action and therefore of cognition as well. In chapter 7 we propose a set of concepts for conceptualizing issues of learning science from the perspective of the unknown, as an encounter of the learner with the (radically) foreign/strange. We draw on ethnographic work in one undergraduate physics course at a Canadian university, where we followed in particular one female Japanese student, who had come to this country for the purpose of getting a degree. As an entry point and as source of empirical materials, we draw on our auto/ethnographic experience as immigrants because doing so comes with particular advantages to ally pathos (the capacity to be affected) to the experience of the foreign/strange. We view this experience as something that is happening to (affecting) us; and this something is beyond all prior experience, understanding, and anticipation. We articulate three phenomenological dimensions that pathos (empathy) allows us to understand concerning the experience of the foreign/strange and then provide an exemplary and exemplifying analysis.

In chapter 8 we develop the notion of the body as the ground of solidarity and responsibility. Qualitative research designed to develop ways of understanding and explaining lived experience of human beings is a reflexive human endeavor. It is reflexive in that researchers come to better understand themselves in the attempt to better understand their participants. Consequently, research ethics itself becomes an ethical project, for it pertains to participant and researcher at the same time: Both are subjects, knower and known. Particularly in case of research on learning, reflexivity arises from the fact that the research constitutes learning about learning. How is ethics in research on learning reflexive of ongoing events and changes of the human learning? In this chapter, from our experience of conducting a project designed to inquire into "learning in unfamiliar environments," we develop pertinent ethical issues. First, ethics is an ongoing historical event; second, ethics is based on the communicative praxis of sensuous bodies; and third, ethics involves the creation of new communicative forms. We conclude that ethics is grounded in a fundamental answerability of human beings for their actions, which requires communicative action that itself is a dialectical process in opening up possibilities for acting in an answerable manner.

ISSUES OF METHOD

This book articulates the dynamics of how learners encounter and experience science and mathematics in everyday settings. Each chapter draws on case exemplars to exemplify theoretical claims on sense making, which we have extracted from a series of research projects. First, the examples of learning geometry presented in chapters 1, 4, and 5 are part of a mathematics education research project that studied children's understanding of three-dimensional geometry concepts in the mathematics classroom of a Canadian elementary school. The curriculum consisted of 15 geometry lessons conducted by a university research team in collaboration with a schoolteacher over a period of three weeks. Second, the case examples in chapters 2, 3, 6, 7, and 8 are extracted from a research project on learning physics, which was conducted as part of a larger-scale project on crossing the boundaries between very different activity forms and the forms of cultural knowledgeability required in each. The purpose of the project was to study students' learning and development of identity across different cultural settings, particularly in the areas of physics at the university level.

The data sources exist in the form of recorded Hi-8 digital videotapes, curriculum materials and students' worksheets, teachers' handouts, and researchers' field notes. The video recording of the elementary mathematics consists of a series of geometry lessons conducted in the school classroom. The video recordings of university physics covered curricular classroom activities (lectures and lab experiments), consultation and problem solving sessions (assignments), and non-curricular university sessions (e.g., department seminars). We digitized the videotapes into QuickTime files. The digitized files were segmented and transcribed. Offprint images were created from the video files and imported to transcripts, which ultimately were replaced by hand-drawings. We saved sound tracks of the episodes as separate audio files. We used a freely available, multi-platform software package PRAAT (downloaded from http://www.praat.org) to analyze prosodic properties of utterances such as pitch, pitch contours, speech intensity, and speech rate.

We maintained prolonged engagement, persistent observation, and peer debriefing with disinterested colleagues in the research laboratory to meet the trustworthiness criteria (Guba & Lincoln, 1993) throughout the research. For example, the participants in the research on learning physics were undergraduate physics students and lecturers. One of the research participants was an international female student (pseudonym Mariko) studying physics. To gain a better understanding of the learning environment in the way it is salient and relevant to learner's everyday life, we conducted intensive ethnographic work involving our participation in various settings of learning physics and collecting a wide range of data sources. We conducted research observations and associated interviews about her physics study, which were followed by videotaping two undergraduate physics courses (thermodynamics and optics) and other school practice over a semester. Through our prolonged study of Mariko, we came to understand the gap between her efforts and positive comments on physics lessons on the one hand and her real performances on the other hand. Our analytic approach offered a possible explanation of the gap (chapters 2 and 6) and of participative (unindifferent) thinking (chapters 3, 7, and 8). This approach does not attribute characteristic phenomena to the capacity of an individual (student, teacher) or the nature of discipline (mathematics, physics) isolated from the concrete situations of teaching and learning but respects the aspects of the situation publicly made available to participants. This method increases the possibility to develop a theoretical framework that is transferrable and testable and also simultaneously deepens the understanding of the particularity of the lecture situation. The authenticity criteria have also been attended to. For example, we watched some of the videotaped lessons with the key research participant. In our collaborative session with Mariko, we observed that she developed a better understanding of physics concepts when she had an opportunity to watch a lecture-video and see/listen to sense making resources deployed in the lecture in the forms other than her own notes.

We also developed a reflexive form of research. The researcher and the research participant came to constitute the reflexive relation between the two and afforded two important aspects during the fieldwork: On the one hand, it provided discursive resources for establishing a rapport and facilitated discursive interaction between the two, and, on the other hand, it provided the researcher with a reflexive ground for taking a first-person perspective on the research participant's experiences refracted through the lens of her learning experiences. In chapters 7 and 8, the reflexive relation between learning to do science and mathematics and learning to do research on it provided opportunities to experience moments of research activities explicitly through an ethical (rather than metaphysically cognitive) lens. It also allowed us to bring the emerging ethical issues to bear on the research. In this sense, these studies embodied a dialectical process in which our understanding of ethics evolved. Each step constituted research praxis producing a description of ethics as lived praxis and, by the very praxis, became a ground of reflexive step. In chapter 8, we see a moment that the researcher had experienced with the research participant (episode), which brought forth an ethically critical question for the researcher herself (SWH). The moment became a constituting event in her learning to do research *and* in research on learning. This was a first reflexive step that transformed the lived experience into the description of the lifeworld. In the next step, we came to provide a description of the issue having changed as we discussed it in a reflexive fashion, that is, reflexive to the process of raising and phrasing the question. This is the phenomenological method Georg W. F. Hegel (1977) used in *Phenomenology of Spirit*, where the outcome of the study is an understanding of the process of arriving at the outcome. Consciousness comes to understand itself as an instance (concrete realization) of collective but contingent consciousness, which means that the process of arriving at this understanding is itself contingent.

In the course of this book, we employ the following transcription conventions:

[and]	Square brackets in consecutive lines with or without numbering show an overlap of speech with body movements, which extends for the duration of speaking the words underlined;
((*draws*))	Italicized words within double parentheses constitute transcribers' comments on visible body movements;
*	Asterisk mark denotes an instant that corresponds to a drawing of video-offprint of which the figure number is labeled at the end of the comments;
(2.0)	Number within parentheses indicates elapsed time of pause in tenth of a second;
NOW	Capitalization marks speech that is louder than the normal speech intensity;
<<p>well>	Words within angle brackets indicate lower speech volume (pianissimo) than normal;
(find out?)	Question mark in parentheses indicates inaudible utterance(s);
cu:be	Lengthening of a phoneme is indicated by colon;
?,;.	Punctuation marks are used to indicate characteristics of speech production rather than grammatical units;
it's-	Dash indicates sudden stop of talk;
=	Equal sign at the end of one turn and at the beginning of the next indicates latching turns, that is, there is no gap between the two speakers;
↑↓	Arrows indicate a rise or fall in intonation sharper and more clearly noticeable than normally occurs.

PART A

FROM THE SENSE OF THE BODY TO THE BODY OF SENSE

> Generally speaking, the new psychology has revealed man to us not as an understanding that constructs the world but as a being thrown therein and attached to it by a natural bond. As a result it re-educates us in how to see this world that we are in contact with at every point of our being, whereas classical psychology abandoned the lived world for the one which scientific intelligence succeeded in constructing. (Merleau-Ponty, 1966/1996, p. 68)

This introductory quote derives from a text on cinema and the new psychology, which Merleau-Ponty found in Gestalt psychology. The author makes a point fundamental to the current book project. Human beings are *not* constructing the world but are thrown into it, being connected to it "by a natural bond." We live in a real world, together with our thoughts, not in a netherworld that is the result of mental constructions. Despite the many problems others have shown with respect to the constructivist metaphor, many science and mathematics educators continue to hold on to it. In this section particularly and throughout this book generally, we mobilize evidence for the central role of the body in the way that Merleau-Ponty conceived it.

Learning science or mathematics involves students' exposure to different sources of sense experience, some of which students have never come across before. For example, in science classrooms students come into contact with visual graphics like the double helix structure of the DNA molecule, sound-words like the names of chemical elements in the periodic table that are not part of everyday discourse (molybdenum, rubidium, or strontium), or equipment used in laboratories for the first time through the experience of seeing, hearing, and touching. Central to this occurrence is the double nature of sense experience. First, the contact with objects from the cultures of science and mathematics consists of bodily and sensuous (visual, aural, and tactile) experience. To perceive some new aspects of the world, students have to let their bodies affected by the different sources of sense (sensuous) experience. Second, at the same time, the sensuous experience of the world does not remain at the material level but provokes sense that constitutes a form of knowing the world. That is, students perceive the world by letting their sense experiences mark some sense on them, without which learning would not be possible. Coming to see the structure of the DNA molecule while seeing the complicated image of the double helix and hearing a teacher's verbal explanation involves this double nature of sense experience. Phenomenological philosophers such as Maurice Merleau-Ponty have explained the double nature of sense experience in terms of the living body. This body is different from a machine that simply collects information from outside. This is not the body studied by biologists, who do not focus on the *living* part but only on the way life manifests itself to the observing human being (Sheets-Johnstone, 2009). The observer's body is not just a material entity but a form and instantiation of life that is capable of acting and at the same time being affected and self-affected by the act that it produces. Thus, the expertise in recognizing DNA patterns while gazing at complicated visuals in the film would not be possible without the living body that inscribes (structures) itself in and through tedious bodily work (e.g., eye movement) and learns how to see.

Therefore, the phenomenological notion of the living body provides a clue about the development of thinking. For example, when someone tries to talk about new ideas and do not find appropriate words, we may observe the person's body engaged in looking for them such as with hands leafing through notes and eyes gazing at diagrams. The lived (bodily) experience of the world is not transparent but "thick" in the sense that the conscious mind cannot objectify the whole – if this were to be the case, the mind would no longer be engaged in learning something new. The living body constitutes the hub of sense experience, on the one hand, and the hub of scientific and mathematical sense making, on the other hand. This double nature of the sense experience constitutes the source of scientific and mathematical perception and learning in and through everyday occurrences of classroom communication. Vygotsky already acknowledged this double nature of sense making experience when he proposed that verbal thought (or verbal thinking) constitutes the whole unit to be investigated to study the child's scientific concept development rather than language or thought alone. In *Thought and Language* (Vygotsky, 1934/1986), the author suggests that we ought to be studying "the development, the functioning, and the structure of this unit [i.e., verbal thought], which contains thought and speech interrelated" (p. 6) rather than "breaking it up to its component elements, thought and word, neither which, taken separately, possesses the properties of the whole" (p. 211).

In this first part of the book, we study the double role of the living body in knowing and show the great power of this idea to explain everyday practice of teaching and learning. We follow and expand the holistic framework Vygotsky presented along dialectic, phenomenological traditions. We show that perceptual experience and sensuous experience – two moments of sense – constitute one another in and through the living body. That is, we theorize that learning means coming under the influence of various forms of sense experience and perceiving them as totalities. The material bodies in the classroom become sense-making resources only within the living (sensuous) praxis of doing (talking) science or mathematics where the living body constitutes the organizing center. The living body with its senses comes to constitute the body of sense. We theorize and exemplify this general principle in two different contexts of teaching and learning concepts, which are often considered abstract and belonging to a realm separate (independent) from actual, sensuous praxis. In chapter 1, we study a mathematics lesson in which little children learn about three-dimensional geometrical objects and associated concepts. In chapter 2, we investigate a university physics lecture in which a professor talks about thermodynamics concepts to physics undergraduate students. In our analysis of teaching and learning at the two different levels of teaching and learning, we exemplify the corporeal and incarnate nature of real-time thinking and conceptual understanding. In both cases, we show that the living body captures the very "more" of teaching and learning that (mechanical) information processing cannot capture; that is, the living body constitutes both the venue where the conceptual gain is achieved and the substance of conceptual again itself.

CHAPTER 1

MATHEMATICS IN THE FLESH

> This displacement of the memory from the domain of thought to that of the flesh, this corporeal memory of which Maine de Biran had the unheard-of intuition, doubles itself depending on whether it is considered at work in the delivery to the senses or in its immanence, before the intervention of any intentionality. (Henry, 2000, p. 207)
>
> Biran did not reduce consciousness to motility but he identified motility and consciousness. . . . It is not a consciousness becoming movement, but a consciousness reverberating in movements. (Merleau-Ponty, 2001, p. 64)

Memory is probably the single-most important phenomenon to be attributed to the capacities of the mind. Yet as this introductory quote shows, there are forms of memory that are better thought of as corporeal rather than mental phenomena. Intentionality and our fundamental capacities related to agency are built on such capacities that predate mental representation. Although we have begun to develop a theoretical framework that takes serious the phenomenological critique of intentionality and agency (e.g., Roth, 2010a, 2011a), a lot of work remains until a more comprehensive theory is developed. The present chapter contributes to this work of a phenomenological account of mathematics (science) in the flesh.

In this chapter, our intention is to articulate a more comprehensive approach to development than exists in Vygotsky's framework and its logocentric approach (he restricts himself to words and "word-meaning"). That is, this chapter takes Vygotsky's theory of participative (unindifferent) understanding and develops it to include not only words but also the living body as a whole, which has many other means for articulating sense than the vocal cords alone, to constitute a more holistic approach. We draw on phenomenology to propose a different way of understanding the nature of mathematical concepts. We substantiate the theoretical issues by presenting exemplary case studies that exhibit children's trajectory of learning geometry in a second-grade mathematics classroom. We exemplify an instance of verbal thinking in which a child engages in knowing the world (object) in and through the living body. We conclude that the extended framework of participative (unindifferent) understanding allows theorizing the role of emotion in knowing and learning mathematics without separating it from cognition.

INTRODUCTION

The "construction of meaning" tends to be the main pedagogical goal for the teaching of mathematics. Yet, the metaphor of *construction*, which is also used to theorize conceptual development and conceptual change, describes learning by taking recourse to that which is unknown to students and therefore produces some irresolvable problem to explaining learning, as it is lively realized. For example, imagine a person who does not know (see) triangles. How can the person *aim at* (intend to) learning (about) triangles if they are both unknown and invisible to the person and from within the world of the person? Because the person is blind to this aspect of the world, and because there is no concept (however defined) of triangles, how can this person learn (communicate) about triangles? Some readers might too rapidly suggest, well the children can touch, or adults can lead the child. But the research in Russian institutes for deaf-blind children shows that much more is required for these children to learn, for when they arrived in these institutions, these children were in mere vegetative states (Meshcheryakov, 1979). They did not have "innate" exploratory instincts in the way that Jean Piaget or Ernst von Glasersfeld postulated. There is therefore a real learning paradox that I cannot intend what I do not know, and, as shown for the deaf-blind, particular conditions are required to get me of this double bind. This is the essential idea of the learning paradox that constructivist educators have not resolved. That is, because students do not know a concept, they cannot at the same time aim at (intend to) learning it; if they knew the concept, necessary for being able to make it the object of an intention, then they would no longer need to learn. Even social constructivism has not solved this problem, because if individuals cannot see and do not understand triangles, they cannot do so in the social interaction.

Lev Vygotsky is the key person in psychology who opened a new path to solving the problem by attending to the inter-functional relation between thought and speech in the child development. In his book *Thought and Language*, Vygotsky (1934/1986) studies the roots of thought and language and suggests that at some point of the child development the two lines of development encounter of which the central phenomenon is participative (unindifferent) understanding. The signification of the word is not a stable property belonging to the word (Bakhtine [Volochinov], 1977). Participative (unindifferent) understanding consists of an ongoing *living* process that develops as a thought is verbally articulated in speech and also affected by speaking simultaneously. In fact, speaking and thinking are not stable entities but processes, and the overarching phenomenon that gathers them into the same unit is itself a process. "Participative (unindifferent) understanding" is a process at three levels: it develops from moment to moment, in the course of a person's change (ontogeny), and throughout cultural history. Thus, to study the child's concept development, Vygotsky highlights the significance of considering "the full sense of the term" in the fullness of communication. The very possibility for the development of participative (unindifferent) understanding is involved in the whole unit of communication in which the sound-word is mobilized and marks sense in some way. Vygotsky's holistic approach to the participative (unindifferent) understanding reveals some mechanism by which mathematical concepts first come to mark sense for the learning students. For example, if learning is an intentional (constructive) process, it has to be framed in such a way that it is entirely located in and contextualized by students' everyday experiences and language. For students, everyday experiences and concepts therefore constitute the very condition that students have to depend on when they learn about it. For students, the conceptual growth involved in mathematical conceptions pertains to the transformation of these mundane experiences and everyday (spontaneous) concepts as the material to be transformed. This material is given in its everydayness rather than by an act of the conscious mind that somehow constructs an object that it does not know. Philosophers have recognized this as the fundamental problem of all forms of constructivism: How can the mind know that what it does and thinks relates to anything outside the body. Even the radical constructivist "solution" that the matter is not one of truth but one of fit does not get us out of the problem, for the constructivist mind is so caught up with itself that it cannot go outside to check its "construction" for fit. The constructivist mind knows no outside (Henry, 2003).

Vygotsky's holistic approach is in line with the phenomenological idea that the living body has the capacity to open itself up to be touched by the world, which deals with the learning paradox somewhat successfully. In the sociology of Pierre Bourdieu (1997), it is precisely this capacity of the body to be open to the world where it is affected that leads to the structural homology between the social and material world and the individual's dispositions for seeing and acting. The holistic approach to sense experience follows a Marxist way of thinking, materialist dialectics (or dialectical materialism), which is a framework that integrates the material and the ideal (consciousness, thought) within a single unit of experience. Although the material and ideal are different, they are recognized as manifestations of a higher unit – as wave and particle are recognized today as different manifestations of light. Yet, there are limitations in Vygotsky's logocentric approach to language and communication. Although Vygotsky points out the significance of analyzing "whole unit," he analyzes participative (unindifferent) understanding only by leaving out the body, which is also the way Vygotsky's work has been adopted in many educational studies after him. The purpose of this chapter is to bring together Vygotsky's cultural-historical perspective on thinking and speaking with Merleau-Ponty's analysis of the role of the body in communication. The living body solves the learning paradox because the corporeal engagement in/with the world not only transforms the world (objects) but also brings about changes in the living body, which is structured in/through the experience. The living body that senses and is affected by the world constitutes the integral aspect of learning in which the invisible/unknown becomes visible/known to the person. Therefore, we theorize the central aspect of the living body in learning mathematics (science) and propose a more comprehensive approach to concept development (formation). Our theoretical extension in this chapter bears two-fold significance. First, we extend Vygotsky's framework of participative (unindifferent) understanding to include the living body that has become an important dimension in educational research in general and research on mathematics education in particular. Second, the extended Vygotskian framework contributes to resolving the learning paradox and the problem of constructivism.

FROM WORDS TO THE LIVING BODY

Communication is central to the development of participative (unindifferent) understanding in learning mathematics. Listening to the other talking about and interpreting representations allows students to participate in the classroom communication and come to encounter conceptual resources new to them. For Vygotsky (1934/1986), the development of participative (unindifferent) understanding in communication constitutes a central phenomenon that explains the development of scientific concepts and the role of instruction. The significance of communication as linked to the conception development is that it allows us to attend to all the extra-linguistic capacities that are part of any practice. For example, gestures and the multimodality of communication have recently gained a lot of attention in mathematics education. That is, anything linguistic in language use therefore bottoms out in forms of experiences that are pre- and extra-linguistic. In this framework, learning mathematics is like learning a language that is part of a larger unit encompassing the fullness of life – all the bodily and corporeal resources that bring about the living/lived experience of the world are also related to conceptual development.

The Living Body as Expression

In his introduction to English version of Merleau-Ponty's (1969) *La prose du monde* [The Prose of the World], the translator John O'Neill characterizes one of the fundamental messages of the book in this way: "we are the language we are talking about. That is, we are the ground of language through our body. It is through our body that we can speak of the world, because the world in turn speaks to us through the body" (Merleau-Ponty, 1973, p. xxxiii). That is, language, our living bodies, and the world are intertwined, flesh of the same flesh. Conceptions, always already expressed through my body and language in a familiar world, therefore have to be thought in this same manner.

Conceptions have origins and trajectories in child development. Mathematical conceptions in this chapter are related to the temporal development of participative (unindifferent) understanding in communication rather than "meaning" that is assumed to reside in specific forms of representation (e.g., words, mathematical representations, and other forms of scientific representations). Our approach to the development of participative (unindifferent) understanding follows dialectical theories of thinking and speaking and its extension to a body-centered comprehensive framework. First, we draw on theories of thought and language that is part of a larger unit encompassing the fullness of life (Bakhtine, 1993). Participative (unindifferent) understanding arises from the process that integrates thought and speech (gesture) dialectically – "continual movement back and forth from thought to word and from word to thought" (Vygotsky, 1934/1986, p. 218). Second, we follow other studies that have extended Bakhtin's participative (unindifferent) understanding dialectic toward the point that thought is dynamically related to the whole unit of communication rather than to words alone. For example, in communication, words take forms of sounds (e.g., prosody) and constitute one part of the whole network that marks sense (i.e., living participative [unindifferent] understanding) together with other corporeal forms of experiences mobilized simultaneously (e.g., gestures [McNeill, 2002]).

The extension following the Vygotsky-McNeill approach makes two significant contributions that lead to a more comprehensive approach to conceptual development than exists in Vygotsky's framework (i.e. participative (unindifferent) understanding). First, conceptions – the concrete ways in which concepts are realized in and by individuals – are distributed across many different forms of experiences, language, gesture, and body (the whole, including emotions). That is, rather than consisting of words alone, we understand "conceptions" to be grounded in the experience of dwelling in a world so that our entire body becomes a source of expression (Merleau-Ponty, 1945). In this move, we take into account my everyday ways of being, for example of talking to others, where I do not experience myself as a computer that mobilizes conceptual structures and concepts stored in my mind. I just talk in the same way that I move my legs and feet: I simply walk. Moreover, in communication I do so as an entire being, and anything I do may turn out to be a general expression that tells others more than what I intend. From my bodily turns or intonations, others take that I am disinterested or angry even though I might not have had such an experience or, I learn through these expressions that something does not interest me and that I am angry. Both Vygotsky and Merleau-Ponty point out that when I speak, I find out what I am thinking: There is not first thought, which is then emptied out into the public. Speakers find out about their thinking as much as listeners. This more holistic theory of conceptions considers different, irreducible modes of communication as a whole.

Second, in this framework of communication, conceptions then may express themselves concretely as part of an embodied life that is irreducibly interconnected with language; therefore, everyday lived experiences constitute the condition for conceptual development. This focuses our attention on the processes by means of which everyday non-scientific conceptions come to be transformed into scientific conceptions. Rather than being eliminated and eradicated, everyday conceptions are the ground, material, and even tools in a transformative process that leads to scientific conceptions. To understand learning, we need to understand is this transformation. We must not theorize learning in terms of the abandonment ("eradication") of old conceptions, because this would also mean that we have to think of abandoning (eradicating) all the tools we have available, in fact, abandoning the very ground in which any thinking occurs.

In this way proposed, the theory of participative (unindifferent) understanding includes the dynamic role of the living body, which allows, for example, emotion to become an inner part of thinking in the way Vygotsky asked for. The following four points summarize the dynamic of development in communication from an extended Vygotskian, body-centered perspective. First, the living body constitutes the mediating hub in experiencing the world (objects). Second, the living body constitutes the mediating hub in communication; my body *is* my expression rather than merely a tool for expressing what is in my mind. Third, the real-time articulation of thinking with and for the other is distributed within the unit produced by the bodily action in itself and with respect to them (e.g., speech, gestures, eye gaze, body orientation and movement, etc). Fourth, eye gaze, gestures, body orientation and movement, which are involved in experiencing the world (objects), are also involved in communication. The four principles explain the integral role of the living body in the translation of the whole unit of participative (unindifferent) understanding in which the living body mediates the unity of different forms of experience: "The unity and identity of the tactile phenomenon do not realize themselves through any synthesis of recognition in the concept, they are founded upon the unity and identity of the body as a synergic ensemble" (Merleau-Ponty, 1945, p. 366). The role of the living/lived body guides the analysis of development without having to mystify it or begin from the dichotomy between spontaneous and scientific concepts.

Concept Development at Three Levels

The proposed comprehensive model considers the development (translation) of mathematical concepts (participative [unindifferent] understanding) at three levels: the cultural-historical level (e.g., geometry as a field of study), the ontogenetic level (e.g., the child development), and the situational level (i.e., concrete events in a mathematics lesson). The trajectory of a child's conceptual development involves dialectically related temporal scales. First, conceptual development involves an individual's participation in the reproduction of participative (unindifferent) understanding that has been established and develops at the cultural-historical level. For example, children in the mathematics classroom listen to a teacher explaining geometrical shapes of three-dimensional objects using different sound-words (e.g., cube, sphere, cylinder, prisms, and pyramids). For children, those sound-words may be terms that are not used in their everyday talk and therefore constitute foreign sounds that they are confronted with in the mathematics class. However, words used in geometry have a long history traced back to the ancient Greek in which people at that time used terms that emerged and grew out of their everyday life and experience and have been transformed over a long history, of which the trajectories are left in etymology. Here, the cultural-historical development is tied to the ontogenesis of individuals who became the names associated with particular concepts. Culture developed as, for example, Euclid, Pythagoras, and Thales worked on the statements that we know today under their names. But the same process occurs today, for observed changes in the culture of science and mathematics occur only when the "driver" of change is in the performance of culture itself. Culture is performed precisely when children engage in relations with their teachers and the cultural artifacts in their classrooms. Because conceptions and (mathematical) literacy require the mobilization of a network in which sound-words make links to other forms of experience, including those that make our everyday life (Roth & Thom, 2009a), we have to conduct our analyses simultaneously at the individual developmental ("Where is the student in her development?") and cultural-historical levels ("Where is the [pedagogical] culture at this point in time?").

Second, the development of participative (unindifferent) understanding at the ontogenetic individual level occurs as the individual actively engages in the salient aspects of the world (objects). Here we have to be cautious, for what is salient to an adult teacher is likely not the same than what is salient to a child, who is only in the process of learning geometry. These different ways of seeing and experiences come together in the classroom, where all participants take for granted that they are in the same world, even though closer analysis shows that they are acting in the different worlds given to their perceptions. The here-and-now of the situation constitutes the setting in which corporeal forms of experience are mobilized and where links are made in appropriate ways. Children learn geometrical concepts by participating in concrete situations in which they have to point out specific objects or speak this or that – which we show in the next section. That is, the ontogenesis of scientific concepts is tied to the microgenesis of children's talk as it unfolds in real time. Therefore, the consideration of the development of participative (unindifferent) understanding at the three levels – cultural, individual, microgenetic – and their intertwined relations allows us (a) to integrate embodied and bodily forms of knowing into the unit of participative (unindifferent) understanding and (b) to analyze a child's conceptual development from a holistic perspective.

SENSE EXPERIENCE AND MATHEMATICAL CONCEPTIONS

Conceptual development occurs over time. In fact, from a phenomenological perspective, we ought to say that *developing*, as process, means *producing time and space*. Because this idea is difficult to comprehend, *writing* (Fr. *écriture*) has been proposed as a metaphor (Derrida, 1967), for in writing, as the pen moves across the page, both space and time are produced, and, with it, words and ideas unfold and develop. The child in a geometry classroom participates in talking about three-dimensional objects and temporally develops its understanding of concepts. In a way very similar to writing, the moving body expresses by creating space and time of communication. The act of speaking or listening to others' talk unfolds in time as much as it is making time. That is, time is generated as the conceptual possibilities that classroom objects make available are realized into different forms of experience. Certain ways of being-in-the-world emerge as the living body temporally engages in objects and therefore spatially realizes different forms of experiencing the world. Those (temporally emergent) different forms of experience are all potential forms of knowing the world that have a significant role in the child's conceptual development. They involve the potential to affect the real-time translation of the whole participative (unindifferent) understanding unit and therefore they are simultaneously unique and partial representations of a higher communicative unit. For example, speech and gestures are two irreducible components both of which dynamically incorporate the context of communication in an integral way and therefore affect the development of the contents of communication. Each constitutes a potential opportunity from which a new way of knowing the world (as a result of the translation of the previous) emerges and begins to grow. Therefore, the child's conceptual development pertains to the dynamic coordination of different forms of experience. Any one form of experience belonging to the emerging concept can activate any other form. Any one form of experience therefore can stand for the totality of experiences that makes the concept. Such a relation, where a part comes to signify the whole of which it is a part, is known as a metonym (Roth, 2011a). In this way of thinking about concepts, therefore, every experience is part of the concept. That is, whereas in other theoretical approaches children's interaction with concrete materials is merely a stepping-stone, kicked away once the abstract concept is attained, in our approach a concept implies all the practical experiences a person has had and that leave traces in the living body. In this manner knowing a concept never is severed from "applying" a concept such that one cannot know a concept without also knowing how to apply it.

The living body is central to this spatiotemporal coordination because it mobilizes different forms and more so makes links between them. The unity that the living body makes available allows the constitution of a participative (unindifferent) understanding unit from which a higher-order cognitive function arises. In this section, we conduct a case analysis and exemplify the role of body in the spatiotemporal translation of a participative (unindifferent) understanding unit. We show how a child bodily engages in the world and thereby develops an understanding of three-dimensional geometry concepts. In the following excerpt from a geometry lesson, second-grade children participate in identifying a mystery object placed on the glass panel of an overhead projector. The object, which is surrounded by a wall from paper, is itself invisible. All that the students see is a shadow projected on a screen and different three-dimensional objects on a shelf below a chalkboard. Therefore, students are given opportunities to talk about the geometrical shapes of unknown three-dimensional objects and their relation to their observed two-dimensional projections. We exemplify a comprehensive approach to conceptual development by substantiating the three-level analysis of the development of participative (unindifferent) understanding.

Episode 1.1

```
01  Teacher:  Clara
02  Clara:    [um]
              [((Clara puts her hand down and stands up))]
03            [(1.8)]
              [((Clara walks to the front))]
04            [I don't think it can be a circle]
              [((Clara grabs a yellow sphere on the shelf and
              turns toward the teacher and other students. *She
              holds the sphere with her fingertips propping
              around the round surface/edge)) (Figure 1.1a)
05            (1.4)
06  Teacher:  [(?? ?) circle?]                  [circles are flat]
              [((Teacher points at Clara))]     [((Teacher holds the
                                                palms of her hands
                                                facing one another
                                                and *moves them
                                                closer))] (Figure
                                                1.1b)
```

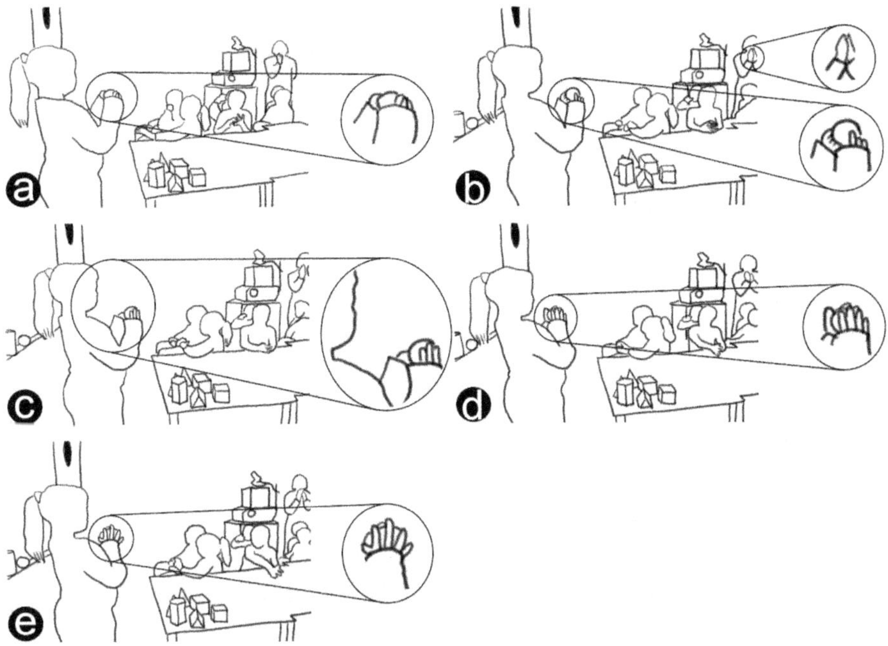

Figure 1.1. Clara talks in front of the classroom a. She holds a yellow sphere with her fingertips propping around the round surface/edge. b. The teacher holds the palms of her hands facing one another and moves them closer. c. Clara gazes at the yellow sphere that she holds. d. She puts the palm of her right hand attached to the surface of the sphere and grabs it by using the whole hand instead. e. She rubs her right palm on the sphere.

07 ((**Clara gazes at the yellow sphere that she holds*)) (Figure 1.1c)
08 Student1: sphere
09 [(<u>1.6</u>)]
[((**Clara put her palms to the surface of the sphere and holds the sphere using her whole hands instead of fingertips*))] (Figure 1.1d)
10 Student2: sphere
11 Clara: sphere↓ ((**Clara rubs her right palm on the right area of the sphere*)) (Figure 1.1e)
12 Teacher: sphere↑ yes, that's right and why don't (you?) think it could be the sphere

Description

Clara raises her hand and the teacher calls her name (line 01). Clara says "um" and walks to the front side of the classroom (line 02). She picks up a yellow sphere, one of the objects placed on the shelf underneath the chalkboard (line 04). She grabs it by using her fingertips and turns to face the teacher and other students (line 04). She holds the yellow sphere as high as her chest and gazes at it. Simultaneously she articulates for the audience that she does not think "it can be a circle" (line 04). A pause comes about (line 05). The teacher points her right hand at Clara and articulates that circles are flat (line 06). Simultaneously, the teacher puts the two palms of her hands together, which constitutes a gesture of narrowing (line 06). Clara gazes at the yellow sphere in her hands (line 07). One of the students sitting in the classroom utters "sphere" (line 08). Clara puts the palm of her right hand attached to the surface of the sphere and grabs it by using the whole hand instead of fingertips (line 09). Another student utters "sphere" (line 10). Clara repeats the word "sphere" and rubs the right surface of the sphere using the palm of her right (line 11). The teacher repeats the word ("sphere? yes, that's right") and utters "why don't you think it could be the sphere" (line 12).

Analysis

In this episode, we see a child participating in a conversation about a mystery object and verbally thinking by talking to a teacher and other students in this second-grade geometry class. (The episode exemplifies a beginning [formation] of mathematical thinking realized in communication.) Clara proceeds with picking up a sphere and shows it to others. She holds a round surface and edge of the sphere using her fingertips and suggests that she does not think "it can be a circle" (line 04). After a 1.4-second pause, the teacher points at Clara and says that circles are flat. The teacher's action provides concrete form of knowing "circle." The utterance "flat" and her gestures of narrowing the space between her facing palms connects the word "circle" to flatness, which contrasts to the spatial shape of the sphere that Clara holds. Clara's utterance juxtaposes two words, "it" and "circle," neither of which directly refers to the object that she holds up (i.e., a sphere). Therefore, the teacher's action makes Clara's coordination of the utterance and the act of picking up the object problematic: It is not clear why she picks up a sphere among others and what she means by "it" or "circle." Clara gazes at the object in her hands. While participating in the talk about the mystery object, Clara encounters something foreign/strange: it makes an appeal and demands a response.

One of her classmates utters "sphere," and thereby makes a sound-word available in the classroom. The articulation of the sound-word "sphere" opens an opportunity for the emergence of a different form of experiencing the object. Clara puts her palms to the surface of the sphere, and thereby changes her way of holding the object from using her fingertips to using the whole hands. This allows her to touch the round surface of the sphere rather than the circular edge of the surface. This new form of experience is linked to speech when Clara repeats the sound-word "sphere" following another student's uttering "sphere" in the next. Clara's speech translates the experience of the object into the sound-word, "sphere," which also affects the experience of the object into her rubbing movement over the surface of the sphere using the palm of her right hand. Clara's body mediates the translation between the physical experience of the object and the hearing of the sound-word. Both experiences are corporeal, which means, it is precisely the living body that constitutes the translation. The two experiences are the experiences of the same "I can." Clara's change of her way of bodily holding (experiencing) the object constitutes a point at which she explicitly changes the contents of her speech but also lets a new participative (unindifferent) understanding unit emerge. It makes for an opportunity through which everyday forms of knowing the world (e.g., round surface of a sphere) are mobilized in such a way as to link to the sound-word and other forms of experience (e.g., teacher's gesture). This therefore expands a network of experiences that have left their mark in and on the living, lived body that we identify with the name Clara. From this holistic perspective, conceptions, "emerging in and from experience, exist in and as of experience" (Roth & Thom, 2009a, p. 186). Any single experience serves as an entry point into this network, because it is not only an integral part of the conception but, as a signifier, also stands for the network as a whole (the signified). The relation of any experience to the whole therefore is of metonymic nature.

The episode exemplifies the beginning of mathematical thinking to which the child's basic experiences in her everyday life are integral. For Clara – or anyone else in this second grade mathematics classroom – the round surface of a sphere may not be an experience unique to mathematics (geometry) but common to her everyday life. For example, when children play with a ball, they have opportunities to touch the round surface of a ball. Thus, etymology shows that the Greeks words pertaining to "circle" (*kúklos*) and "sphere" (*sfaîra*) originate from the sound-words referring to "ring" (*kúklos*) and "ball" (*sfaîra*), respectively. That is, for Greek people, speaking the word *sfaîra* immediately mobilized an everyday experience of using a ball and therefore constituted a metonymic relation to a network of corporeal experiences related to this toy. Yet, children today have no clue as to the relation between the geometry words "circle" or "sphere" and their everyday experience. These words initially are foreign. They are associated with geometry classes but not with the general experience outside of school. In fact, the "reading circle" or "story circle" do not involve the kinds of ideal objects that geometry is about. In a way, these words are dead metaphors whereas they have been very much alive for the ancient Greek, for whom they denoted everyday experiences. Today, English-speaking children have to learn to make explicit links between these new and foreign sound-words to their everyday experience (e.g., playing with a ball) in the mathematical classroom. These experiences therefore become constitutive of new ways of knowing, as familiar ways of experiencing kinds of objects come to be associated corporeally: feeling the sphere and hearing the sound-word both happen in and to the same "I can" that characterizes the living/lived body to which we refer by saying and writing "Clara."

In this way, the episode exemplifies the beginning of conceptual development in which Clara learns to differentiate "circle" and "sphere" by actively participating in making a network of everyday experiences. Yet, again, it cannot be the result of her intention to differentiate them. For example, Clara simply says "circle" because it makes sense to her – she raised her hand and volunteered to speak. Articulating the word "circle" brings her to a situation in which she encounters different forms of knowing "circle" and the object (i.e., sphere) that she holds in her hands (e.g., "flat," narrowing gestures, and "sphere"). In this instant, there is no evidence that

Clara knows what or whether she needs to differentiate. She just engages in touching the object that she already holds in her hands (i.e., the palm curved along the round surface of the sphere) and utters the sound-word ("sphere") that her classmate has suggested. These simple actions bring into alignment the resources that make up the participative (unindifferent) understanding unit. Thus, the teacher utters "why don't you think it could be the sphere," which thereby translates Clara's initial claim (i.e. "don't think it can be a circle" [line 04]) into another (i.e. "don't think it could be the sphere [line 12]). The episode exemplifies the central role of the living body in the child's conceptual development: it provides opportunities for metonymically bridging interpretive resources and the everyday experience of the world in two ways. First, the living body mobilizes different forms of experiencing and increases interpretive resources for knowing the world. Second, the living body coordinates different forms of experience and increases conceptual possibilities (e.g., the emergence of higher-order cognitive functions).

TOWARD A HOLISTIC APPROACH TO VERBAL THINKING

In this chapter we exemplify a comprehensive approach to conceptual development (concept formation) that does not presuppose participative (unindifferent) understanding separated from the everyday conditions that make the communication of concepts possible in the first place. We take Bakhtin's theory of participative (unindifferent) understanding and combine it with Vygotsky's approach to the development of thought and language. This leads us to a holistic approach in which bodily forms of knowing and learning constitute the irreducible condition for conceptions – an idea that Vygotsky, because of his exclusive concern for the word, has not articulated in his writing. We exemplify the ways in which a child's body temporally engages with objects located in local spaces and makes links between different forms of experiences. The living body is central to learning geometry because of the capacity to realize everyday forms knowing the world in a specific setting and to translate them into spatiotemporally coordinated mathematical forms. Because of the unity of the corporeal "I can," the different experiences that come from the hands, body, ears, or eyes are already integrated rather than demanding for a (constructivist, computer-like) mind to coordinate them. We suggest that conceptions involve the development of my familiarity with my lifeworld, that is, of my knowing my way of around the world. This development coincides with the development of communication (expression) in situ. In a nutshell, therefore, we can say that as the child's talk unfolds (level 1), the child moves a little bit along its developmental trajectory (level 2), and a bit of mathematical cultural history is produced (level 3). It is precisely such a conception that allowed Bakhtin (1981) to explain the development of the novel (level 2) in the history of Greco-Roman culture (level 3) as people used language in conducting their everyday affairs (level 1).

Sense making is a core aspect of science and mathematics education and generally discussed in terms of language. This chapter informs us precisely about the source of the possibility of making sense: the living body. It is the living body that bridges and translates between interpretive resources and the experience of the world and thereby contributes to the development of higher-order cognitive functions. The child in our example participates in a cultural activity (i.e., geometry curriculum enacted in the elementary mathematics classroom) and simply works with objects (including words) given to her. The alignment of actions spread across her living body emerges as she encounters the words and other forms of objects rather than through the intentional act toward an outcome that would have to be known to her already. The living body draws on and mobilizes *again* the totality of resources that have a long cultural history. In the children's actions, "the sensual illustration of the concepts" "is surreptitiously substituted" by means of the concrete figures and bodies that stand in school lessons for the idealities of geometry (Husserl, 1939, p. 217).

By extension of this chapter, we direct our attention to emotions that constitute an integral part of verbal thinking when the living body is thought as part of scientific and mathematical conceptual development. Language-centric traditions of psychological thinking have treated emotions additional to thinking and therefore belonging to a separate domain of learning (e.g., motivation, attitudes). Our case analysis in this chapter implies the inseparability of emotions from thinking and speaking as a whole and therefore from sense making in the following two ways. First, the living body produces and reproduces emotional valence toward the object in/of communication. Clara volunteers to speak and to express herself to others (teachers, classmates) in the classroom. Simultaneously this opening of herself to the world in and through her actions also gives her an opportunity to listen to others. Clara's series of action involving facing others and attending to the shape of the object in the next shows the continuous production and reproduction of her openness to others. Clara articulates her proposal to others in a public forum and attends to the teacher's talk and gesture. Then, this communication leads Clara to experience the object in her hands, and thereby realizes the teacher's talk as the request to attend to the shape of the object. Although Clara does not initially produce a hypothesis that a mathematician would consider sound, she neither gives up nor loses her orientation (openness) to the object. The living body continuously engages in knowing the object and makes possible the reproduction and transformation of the emotional valence (i.e., her attraction toward the object). Second, sense making emerges from and develops on the empathetic terrain of communication. This is clear from our analysis that the attention to the object is maintained and develops by communication with others. The living body reproduces and transforms the empathetic terrain of communication where Clara touches the object and articulates for others her sense experience. Emotions are an irremediable part of the terrain of learning and this terrain is generated and maintained by the living body. This explains why students come to develop a good sense of topics (i.e., literacy) when they have opportunities for dealing with scientific and mathematical objects within the very setting where they have opportunities for talking about these same objects.

CHAPTER 2

LECTURES AS CORPOREAL PERFORMANCES

Lectures are often conceptualized in terms of information transfer. Science and mathematics professors tend to talk about using lectures to "get information across" and about how much their students do or do not "get" the main points of a lecture. Other scholars, though they come with very different presuppositions, do not abandon the transfer metaphor but focus on the "construct meaning of" what physics professors (lecturers) say and the notes they put on the chalkboard (overhead). In this situation, the conceptualization still occurs in terms of some stuff that goes from lecturer to audience, but whereas in the first case it is a simple transfer, the interpretation and making of participative (unindifferent) understanding are highlighted in the second approach. In either case, the view does not explain, for example, why students indicate to have a clear sense of understanding while they sit in a lecture whereas they subsequently experience failure to understand their own lecture notes or textbooks while preparing for an exam. Based on more than a decade of studies on the corporeal nature of science lectures (e.g., Pozzer-Ardenghi & Roth, 2010), the purpose of this chapter is to articulate and exemplify a different way of understanding physics lectures. We show that there is more to lectures than the teacher/professor talk plus notes. This informational "more" may explain (part of) the gap between students' participative understanding that exists in the situation where they sit in the lecture, on the one hand, and the one where they study for an exam from their lecture notes, on the other hand. Our results suggest that in lectures, concepts are better thought of as heterogeneous performances in which sense arises from the synergistic and irreducible interrelations of different communicative modalities, including gestures, body movements, body positions, and prosody. Each of these modalities is but a one-sided (and therefore partial) manifestation of the communicative whole. These one-sided manifestations cannot be simply added up to yield the whole, because their characteristic particulars only appear *in relation* to all other one-sided manifestations. There is therefore a part–whole relation such that taking out or adding any single manifestation changes the communicative whole and therefore all other parts. For example, without intonation across a word sequence, we do not know what is meant by "This is a chair." It might be a question, an assertion, or an affirmation. The nature of what is meant also changes according to the particular word or words emphasized. Thus, the sequence "THIS is a chair" is heard differently than the sequence "This IS a chair" or "This is a CHAIR." Such changes generally are not recorded in the notes that the students take while attending a lecture, and yet play a crucial role in understanding what is communicated.

INTRODUCTION

Science lectures are about concepts and how to employ them in thinking about physical problems. Science lectures present concepts to students in real-time and therefore constitute a unique conceptual event that is irreducible to textual forms. That is, in science lectures, professors not only produce sounds that are heard as words but also perform other communicative actions: they gesture, draw diagrams, write equations on the chalkboard, or move around different parts of the classroom. In fact, the inner contradiction of the natural sciences is the disjunction between language, on which participative (unindifferent) understanding is based, and their non-linguistic representations that constitute the antithesis of sense (Heidegger, 1954). Sometimes lecturers show demonstrations, and thereby juxtapose yet another set of performances to the other parts of the ongoing lecture. All of these can be understood as resources for making sense. Yet, in the culture of high school and university schooling, there is a tendency among students and even professors to approach physics concepts only in terms of the verbal texts produced (e.g., textbooks, lectures, and lecture notes) – we do note that there are some science (physics) educators concerned with conceptual understanding, but in the bigger picture, most physics teachers and professors do not adhere and teach to conceptual understanding. Underlying this tendency is the assumption that the performative dimension of physics lectures can be reduced to the (disembodied) reproduction of texts and equations and therefore that embodied aspects of communication other than words (e.g., gestures, body orientation) are supplementary such that their sense can be articulated by means of words. In this quite common approach, complex bodily performances are reduced to language and linguistic signs. But that which can be written down is only one (necessary) part of the objectivity of science; the other necessary part is the actual work of producing and working with language and signs (Husserl, 1939; Garfinkel & Sacks, 1986). Whereas this (generally invisible) work is mobilized on the part of the lecturer, it is not likely produced by the students once they study from their lecture notes.

Viewing language as a carrier of participative (unindifferent) understanding and everything as a supplement runs counter to sociocultural theories of communication, according to which language and all other semiotic (sense making) resources such as gestures are subordinate to a higher order communicative unit, of which language and other resources are but one-sided, incomplete expressions. The adjective "one-sided" means that speech and the gestures that are co-produced cannot be reduced to one another: They are unique and partial dimensions of a higher communicative unit that seems to exist in the lecture as a whole. This position is consistent with recent linguistically oriented studies that show that visual representations – which in the social studies of science are denoted by the term "inscriptions" – come to constitute a communicative unit only in relation to the corporeal actions of the lecturer. Our research team conducted a considerable number of studies over the past decade attending to the deployment of scientific concepts in and through communicative performances (e.g., Pozzer-Ardenghi & Roth, 2010). These studies suggest that communication constitutes a core phenomenon of knowing and learning physics and that concepts, if there are such, can only be "transmitted" from one person to another by different communicative means all of which require bodily performances. Therefore, a study of different modes of communication in physics lectures has the potential not only for articulating the dynamic dimension of language and literacy but also for responding to pedagogical issues such as the role of lectures in students' development of scientific concepts. For example, gestures that seem to provide contradictory information from a visual representation at hand are neither redundant nor independent. Rather, they constitute resources that push the development of the ongoing talk. While lecturing, in and through the bodily performances that somehow translate between what students know and the professor's understanding of the field, a physics professor – in addition to teaching the concepts – also comes to develop knowledgeable ways of talking physics for students who do not yet know the physics concepts.

In this chapter we articulate and exemplify a different way of understanding physics lectures: from a body-centered perspective. We summarize in the upcoming two assertions (captured in the section subtitles) what we have learned over a decade of studying gestures and other bodily expressions in science lectures generally and in physics lectures specifically. In so doing, we also introduce science educators at the secondary and post-secondary level to the cultural-linguistic analysis of the performative dimension in physics lectures. We conceive physics concepts as *performances*, and thereby articulate a cultural-historical and body-centered approach to lectures. In using the notion of *performance,* we appropriate the linguistic approach claiming that teaching constitutes a communicative activity and therefore that capturing the concepts made available in physics lectures requires the consideration of all the resources mobilized simultaneously and in the course of time. We explicitly acknowledge the heterogeneity of communicative performances since it addresses the nature of the social world that students experience when they participate in lectures. That is, students do not perceive what might be in the head of the lecturer – what they concretely perceive is his vocal, gestural, and positioned bodily performance of concepts in the here and now of the classroom. As if they were sitting in a lecture held in the language that the lecture is to teach, students, in attending physics lectures, are confronted with the unknown. They do not yet know what they come to encounter in the next instant and what they will know only as a result of their participation. Therefore, the heterogeneity of communicative performances constitutes the structural possibility for conceptual development by which the audience can come to know some concepts. The lecture, even though it uses concept words to talk about concepts unknown to students also has to provide those resources that students already find intelligible. It is out of the double heterogeneity – across the unknown/known language and multiple modalities – that students are enabled to appropriate from a lecture that which is absolutely and unforeseeable to them.

In what follows, we provide our assertions together with the highlights that summarize the current state of the art. We draw on previous studies of lecturing particularly on the topics of gestures, the use of visual representations in scientific communication, and the role of the body in the constitution of communication space. Simultaneously we engage in an extensive data analysis, which means we closely look at the claims reported in the previous studies and draw on them as the lenses for understanding case materials that we have in our data sources. We use examples from an undergraduate physics course in which the professor talks about the laws of thermodynamics and associated concepts (e.g., adiabatic demagnetization). We analyze the professor's communicative actions and the communicative spaces (fields) that the body physically configures in the course of interacting with artifacts and other structures of the setting. We exemplify how different kinds of performances constitute sense-making resources for communicating concepts and that significance arises from the synergistic, irreducible unit of different communicative modalities, including speech, gesture, body position, body orientation, and prosody (e.g., pitch, speech volume, speech rate).

We build up our assertions on lectures by analyzing the (bodily) performances that are publicly made available in and through communication rather than depending on psychological categories or the bodily correspondences we cannot access (e.g., invisible and hypothesized structures of the mind). Therefore, our exemplary case analyses are relevant to studying the cultural (sense-making) structure of communication and the lived work of physics lectures. As part of implications for improving teaching and learning physics, we propose *translation* as a concept for theorizing teaching and learning physics in and through lectures. Our assertions may well explain the lived experience of participating in lectures to teach or learn physics concepts: that is, some answers for questions such as "Why do students have a sense that they understand while they sit in a lecture and find themselves failing to understand while they prepare for an exam with their class notes or textbooks?" or "Why do teachers (lecturers) in the classroom often ask their students to look at them talking rather than copying texts from the chalkboard into their notebooks?"

CONCEPTS – PERFORMANCES IN AND ACROSS FIELDS

Here we articulate and exemplify a different way of understanding physics lectures. In physics lectures, it is common to use inscriptions, that is, external visual representations that appear on paper, computer monitors, and chalkboards including equations, photographs, graphs, and diagrams. Inscriptions constitute important resources for articulating and communicating concepts. Inscriptions ought not to be taken as inherently communicating something but as requiring tremendous effort on the part of lecturers to make their sense salient for students. For example, previous studies show that lecturers spend a substantial amount time of their lecture talk near inscriptions and produce various communicative resources in association with inscriptions such as graphs, photographs, tables of numbers, or maps. Importantly, these studies show, for example, how lecturers change their body position and orientation in predictable ways when their narration changes: A physics professor moves back and forth between two chalkboards where the Aristotelian and Galilean positions on motion are articulated respectively (Roth & Tobin, 1996) or a biologist reorients when her talk changes from being about an inscription (map, photograph) to being about the real world things these inscriptions stand in for (Roth & Lawless, 2002b). The coordination of gestures and speech is so important that in cases of non-alignment, even scientifically trained audiences have difficulties understanding the lecture. Even more interesting, perhaps, is the fact that when two or more science teachers work together for a significant amount of time – e.g., exceeding two or three months – then their gestures, body positions and orientations, and prosody displayed in lectures tend to become alike (Roth et al., 2005). These studies therefore suggest that the proper unit for analyzing concepts comprises not only language and inscriptions but also the movements of lecturers (e.g., gestures) with respect to the geography of the classroom and the relationships between different modes of communication.

The concern of physics lectures in this chapter is related to the development of participative (unindifferent) understanding. For example, speech unfolds in and through time, thereby also producing the temporality of the lecture. A physics professor in a lecture hall temporally develops the concept and, together with it, thought. Lecture time is generated as the pedagogical possibilities that sense-making resources make available are realized in and through concrete (lived) work. In the presence of inscriptions, the lecturer's body moves not only in and for itself but also with respect to things in the settings (inscriptions and other structural resources); and it is in this relation to other things that the sense of any individual expression becomes salient and thereby marks a feature as remarkable. For example, the relative position of the body with respect to visually available representations parses and mediates the space that is associated with an orientation to the inscriptions, on the one hand, and the space in which narratives unfold about the world that the inscriptions denote, on the other hand. The spatial configurations that the body produces by moving about in the classroom are important sense making resources – even in the case that these are not consciously attended to as in the case of prosody – for students attempting to understand physics concepts. Consider the following examples of a university physics lecture in which a 30-plus-year veteran professor draws a graph on the chalkboard and talks about the third law of thermodynamics. This law can be formulated in this way: As a system approaches absolute zero temperature (0 K), its entropy tends to a minimum while remaining greater than 0 (i.e., $S > 0$). The episodes exemplify that physics lectures are corporeal performances, of which a central aspect is found in the different arrangements of communicative modalities that the body performs or deploys in and across physical spaces.

Physics Concepts Take Place in Different Fields of Communication

Physical arrangements in lectures constitute communicative resources: Speakers' distances from and relative body orientations to inscriptions are integral to the sense of what is being communicated – spatial sense making resources help understand physics teaching in a theater-like context. This is so because both lecturers and inscriptions not only occupy physical spaces but also – because they break the spatial symmetry – mark out preferential spaces in the lecture hall associated with

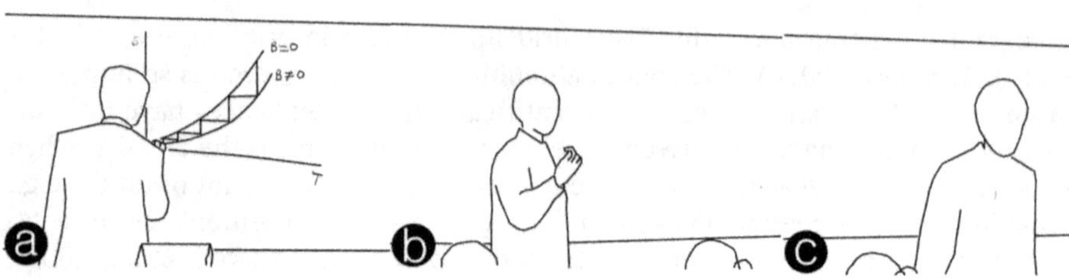

Figure 2.1. a. The professor draws a graph on the board. b. He walks toward the students with his arm shaking. c. He stands near the students on one side of the board.

the sense of what he communicates. The body moves about the hall in lectures, and thereby opens different local fields for the performance of concepts. In fact, the field itself is a contextual constituent of the sense of the concept. In the following episode, the professor is in the process of drawing a temperature–entropy graph of adiabatic demagnetization on which he builds a narrative concerning the third law of thermodynamics. The episode shows that the articulation of the concept in different communicative modes is related to the change of their physical arrangements. Communicative productions take place in different spaces and, therefore, the physics concepts consist of spatially unfolding corporeal performances.

Transcript 2.1

```
01  [and another one]                          [and another one]
    [((draws the fifth horizontal line))]      [((*draws the sixth
                                               vertical line))](Figure
                                               2.1a)
02  [we NOW WE GET VERY VERY CLOSE get]
    [((moves the chalk toward the origin))]
03  [EXTREMELY close to zero]
    [((turns his body to the right and begins to walk))]
04  [but I think this we see from there]
    [((turns his face to the students in walking to the right))]
05  [you will never actually come to get to absolute zero]
    [((*shakes his right arm rhythmically beside his body and
        moves toward the students))](Figure 2.1b)
06  in a finite number of operations  [(2.0)]
                                      [((*moves to the students
                                      and stops))](Figure 2.1c)
07  and that's going to um find an
```

In this situation, the professor stands oriented toward the chalkboard and draws a graph (Figure 2.1a). He draws a horizontal line (the fifth from the right) while uttering "and another one" (line 01). He continues by drawing a vertical line (the sixth from the top) and repeats "and another one" (line 01). The professor's utterance increases in pitch (base speech frequency), becomes louder (higher in intensity), and becomes faster (higher speech rate). He suggests that the stepwise line approaches the origin of the graph and comes "very, very close to" it (line 02). The professor turns to the right and begins to walk. He continues to talk with high speech intensity, suggesting that they get "extremely close to zero" (line 03). The professor walks and turns his face to the students. He makes available to his audience some hesitation ("but I think" [line 04]) and then continues, "we see from there you will never actually get to absolute zero" (lines 04–05) "in a finite number of operations" (line 06, Figure 2.1b). In uttering, he rhythmically moves his right arm (these movements technically are called "beats") in synchrony with the ups and downs in the pitch of his voice; and he turns his step toward the students (line 06). He finishes speaking and moves closer to the students (line 05). The professor stops moving and stands still at one end of the chalkboard (line 06, Figure 2.1c). After a two-second pause (line 06), he continues by uttering: "and that's going to um find" (line 07).

In this episode, we see that the professor's communicative performance take place in terms of different positions and placements in the classroom. Initially, he stands close to the board at the center of the front classroom. He draws straight lines between the two curves on the graph and thereby performs a process of lowering the temperature by means of "adiabatic demagnetization." Here, the movement of the chalk and hand constitute a gesture that metaphorically denotes the physical process. The professor engages with the inscription, and thereby enters what other researchers of gestures in physics have called the "domain space" (Ochs, Gonzales, & Jacoby, 1996). The professor leaves this local space as he finishes drawing. His body turns away from the board and he begins to move to the right, in which he ultimately turns toward the students. Here the professor begins to produce talk of a different genre (line 07). Following previous studies, the former domain space and the new domain space that emerges from the professor's orientation toward the students can be conceptualized respectively as an *inscription space* and a *narrative space*. Very different gestures and narratives are produced when lecturers (professors, teachers, presenters) face the audience (e.g., "you will never actually come to get to absolute zero" and the gestures of shaking his arms besides his body [line 05]) than when they are oriented to the chalkboard (e.g., "we get very, very close" and the movement of the hand and piece of chalk toward the origin of the graph [line 02]).

The differentiation of the narrative space from the inscription space is significant, for the inscription and the physics concepts that it refers to are extended to another local field in the lecture hall but in a different form. Because the two fields are associated with the relative position of the body with respect to the inscription (the visually available graph), on the one hand, and to the audience, on the other hand, the movement of the body constitutes a central aspect of the differentiation and the transition. For example, after finishing the stepwise drawing of lines on the board, the professor turns his body away from the board. More so, there is a rapid increase of the pitch and speech intensity in his utterance around at the instant of turning the body (line 02): Whereas he has been speaking with a pitch in the range from 120–130 Hz, his pitch now moves up to between 282 and 364 Hz; and his speech intensity moves from a mean of 54 dB to a mean of 63 dB, which corresponds to an increase of speech volume by a factor of 8 (each 3 dB interval constitutes a doubling). That is, at this point the speech intensity is eight times of that what it normally is; as the perceptual correlate of intensity is loudness, what the audience members actually hear is indeed a significant increase. (We report intensity, which can be easily mathematized, measured, and quantified, whereas loudness is a perceptual quality that depends on subjective judgments.)

With these substantial increases in pitch level and speech intensity, the professor makes especially salient the point that adiabatic demagnetization never fully reaches absolute zero temperature – the prosodic properties constitute a component that clarifies the concept of the third law of thermodynamics. Whereas one might have been able to see the chalk and hand get to zero, the sudden change in the voice parameters signal attention to what is currently being said, that is, the possibility to get very close to without actually reaching absolute zero temperature (0 K). Here, prosody constitutes a change in sound that the body, through the vocal cords, makes available for marking out the salience and sense. But the body produces these changes without conscious attention so that the difference between the thought expressed in them and the thought itself becomes undecidable, an issue that already Immanuel Kant noted toward the end of his life but could never develop in his articulation of the nature of thought (Nancy, 2008). All of these material resources, together with his walking to the right and the gesture of shaking his right arm, can be seen to constitute a unit that allows getting out of one domain space and into another, which thereby articulates the physics concept, the third law of thermodynamics.

Physics Concepts are Marked by the Heterogeneous Organization of Different Communicative Fields

Inscription space and narrative space are two salient domain spaces (or fields) found in physics lectures. The lecturer's body not only opens up these different spaces but also materially interconnects them, and thereby structures different communicative performances in and across spaces. This is exemplified in the following episode (immediately following the previous episode) where the professor performs an aspect of a concept – the stepwise decrease of entropy – in the form of an analogy. He moves back and forth between the two local fields. From this back and forth movement arises an analogy between forms of performance. That is, the analogy arises from two similar but different performances that are held together by the movement from one to the other space. Because these performances are for the benefit of the audience, the conceptual structures are brought to the surface rather than leaving them – as in other teaching approaches that use analogies – in the depth where they are often inaccessible to the learner.

Transcript 2.2

```
08  other way to [adiabatic magnetization]
                 [((*turns right toward the graph on the board,
                 walks toward it, and stops))](Figure 2.2a)
09  [it's going to (0.9)]
```

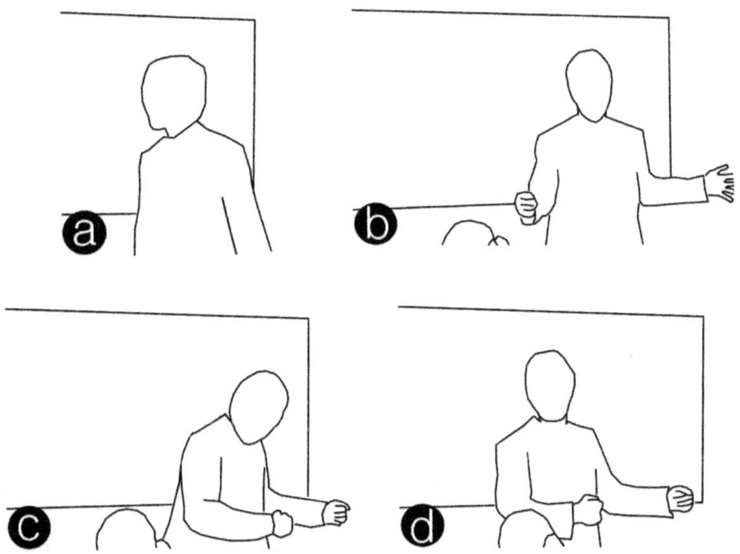

Figure 2.2. a. The professor turns his body toward the board. b. The professor walks back with his arms stretched. c. The professor moves his right arm toward his left. d. The professor returns his right hand to the center of his body.

```
   [((turns to the students, walks to them, and *has his arms
   stretched))](Figure 2.2b)
10 ((breathing)) (0.9) we we we  [talked the other day about]
                                 [((bends his fingers while
                                 stretching the arms))]
11 [isothermal compression]                  [followed by]
   [((*moves right hand toward his left hand))][((stops the
   (Figure 2.2c)                                 movement))]
12 [adiabatic]
   [((pulls his right hand back and stretches to the extent))]
13 [decompression]
   [((*returns his right hand to the center))](Figure 2.2d)
```

The professor continues to talk and thereby engages in the (embodied) performance of concepts in the narrative space. As he utters "adiabatic magnetization" (line 08), the professor's body turns toward the graph (line 08, Figure 2.2a). He takes a few steps toward the board and stops (line 08). He quickly turns toward the students and approaches them with his arms stretched (Figure 2.2b, line 09). He faces the students and utters "it's going to be" (line 09). The utterance stops, and a pause unfolds (line 09). The professor takes a big breath, which is followed by another pause (line 10). He produces words in a mumbling manner (line 10). The professor bends his fingers with the two arms still stretched (line 10). He moves his right arm to his left (line 11). He thereby generates a gesture of "isothermal compression" that he also names (line 11). The right hand stops at the left point (Figure 2.2c) in which he utters "followed by" (line 11). The professor moves the right arm back to the right side of his body saying "adiabatic" (line 12, Figure 2.2d). He utters "decompression" and moves his right arm back to the center (line 13).

In this situation, we see and hear how the communicative resources that are produced while squarely facing the audience in the narrative space (lines 11–13), constitute an analogous structure to those that were produced in the inscription space. The gestures of pushing and pulling together with the utterances of "isothermal compression" and "decompression" (Figures 2.2c, d) can (and have to) be seen as iterations of the stepwise shape of the graph on the board (Figure 2.1a) both physically and metaphorically. The performances in the two local spaces constitute one structural unit for articulating the concepts of "isothermal magnetization" and "adiabatic demagnetization." This is consistent with other observations in physics lectures, which suggests the inference that it is common to present concepts by translating inscriptions to analogous but distinct representations. This episode exemplifies how the body takes a central role in producing those analogous structures by mediating the undecidable relation between communication and engaging in the world.

The preceding paragraphs underscore how physics concepts are constituted in and by means of the heterogeneous organization of different communication fields. On the one hand, the body makes transitional movements and therefore differentiates spaces. At the same time the body produces communicative resources that coordinate those differentiated spaces. For example, in the previous episode the professor stops moving and stands close to the students (line 06). He finishes his transition and, being there, opens up a narrative space (line 07). However, when he begins the narration, he can no longer engage directly with the inscriptions (e.g., visible lines and curves in the graph) that were once available to him. Thus, in the course of saying "adiabatic magnetization," the body turns to the graph, takes a few steps, and comes back quickly (lines 08–09, Figures 2.2a, b). This series of body movement physically connects the narrative space to the graph on the board, and thereby constitutes a *growth point* (McNeill, 2005), that is, the undeveloped and therefore general seed of an idea from which the compression–decompression narrative emerges and develops. The professor orients toward the graph, of which the performance takes place between the inscription space and the narrative space, and therefore mobilizes the graph as an agential component that affects the performance in the narrative space. Conceptual explanation about the dynamic process of lowering the temperature is constituted by the performances distributed over the spatial structure of the classroom.

TEACHING AND LEARNING CONCEPTS IN PHYSICS LECTURES

Lectures are pervasive in high school and university science and mathematics courses. In this chapter we exemplify the embodied performances of concepts in physics lectures and propose a novel approach to understanding lectures as they are heterogeneously realized in the university lecture hall. Many studies on teaching physics (e.g., models of instruction) count on textual (written or verbal) information to analyze what has been taught, and thereby re-present the realities of teaching into disembodied forms. For example, the real classroom events are often attributed to specific teaching/learning models or some descriptions of teachers' intentions and beliefs (i.e., what has been intended or planned to teach). Throughout this chapter, we take an approach that considers the physics professor's labor in the lecture as a singular plural whole. We do not reduce the heterogeneous whole to the part – this is at the heart of a cultural-historical, Vygotskian approach to human knowing and learning. In *Thought and Language* (Vygotsky, 1934/1986) the problem of the reductionist approach to psychology has been highlighted. Our analysis of lectures in this chapter indicates that the communicative value involved in the professor's practical engagement in the world (e.g., inscriptions) exhibits the power of the communication of concepts in physics lectures. The difference in physics lectures from reading a physics book lies in the communicative capacities involved in the professor's lecturing act, that is, in his performative work that his audience comes to witness if not directly but at least in terms of its effects. Therefore, communication constitutes a good paradigm for theorizing science lectures. This allows us to distinguish between students' experiences of participating in the lesson, on the one hand, from their reading the lecture notes, on the other hand.

Multimodality, in the sense that we use the term here, does not mean many modalities that add up to the complete message. Nor does it mean that the same message is produced across multiple modalities. Multimodality, as we use it here, means heterogeneity because different modalities of communication cannot be reduced to each other – for example, the essence of the continuity of a pictorial graph is not rendered by the words "continuity of a graph" or, for these matters, by any other combination of words. We may have the same words – e.g., "isothermal compression" – but very different informational content when the professor emphasizes one rather than another part of the adjective. Thus, saying "ISOthermal compression" draws attention to the "iso," the staying constant, whereas "isoTHERMal compression" draws attention to temperature in a possible contrast to another variable ("isentropic"). Some studies do in fact attend to the multimodal nature of communication in the science classroom (e.g., Kress et al., 2001) – and this may address the social realities of knowing and learning concepts in lectures (particularly the role of students). In these studies, the multiplicity of communication is used to support a new pedagogy to literacy – but the aspects of heterogeneity involved in the notion of multimodality and its implications for teaching and learning concepts have yet to be articulated. Each modality can be understood as part of a whole, the concept (Figure 2.3). Each contributes to the constitution of other modalities, its identity, and, in turn, is constituted by others as well as by itself. The part (e.g., temperature-entropy curve) contributes to constituting the whole and therefore denotes this whole only metonymically, because, as part, its communicative quality depends on the relations within the whole. But, characteristic for metonymy, one part does not capture the essence of other constitutive parts. If we pull only one part or add one new part, then the whole changes and no longer is the same so that, because of the whole–part relation, all the other parts change together with their mutual relations. We could also say that we are dealing with a

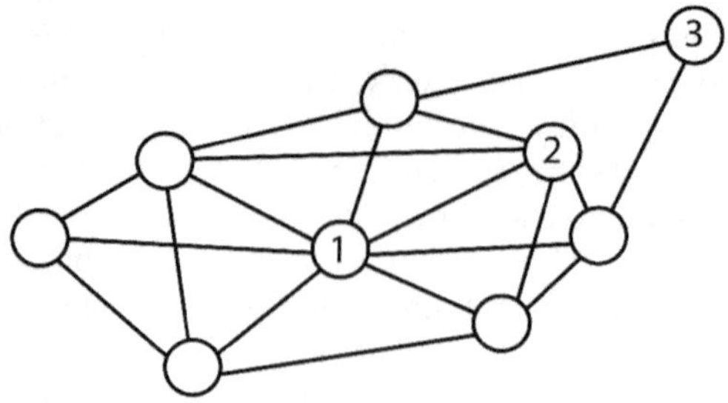

Figure 2.3. An example of a concept understood as a constraint satisfaction network. Pulling node 1 radically changes the network as a whole, including the identity of all remaining nodes, many of which formerly were determined by this node. Also, pulling node 2 or node 3 would change the network in ways very different from pulling node 1.

singular plural phenomenon, where the singular whole really is a plurality of moments, but the nature of each moment depends on the plurality. As a result, in each conceptual unit, the singular is plural and the plural is singular.

Lecture performances exist in time and space. They realize physics concepts by temporally deploying different modalities of communication in and across different domain fields (i.e., inscription, narrative). From a holistic perspective, a central issue related to the two modes is that they are closely intertwined in the course of conceptual development: a movement from concrete to abstract (e.g., mathematical analysis) requires a movement from abstract to concrete (e.g., knowing details of materials at hand). For example, coming to a better understanding of the thermodynamic cycle (i.e., the graph on the chalkboard) goes together with coming to a better understanding of the motion of a bouncing ball (i.e., details of analogy introduced by the professor's narrative) – see chapter 6. Psycholinguistic studies show that heterogeneous communicative productions (e.g., gestures and speech) constitute a dynamic process in which thinking develops all the while being connected to the context of communication – "the intra- and interpersonal planes are inseparable *in all acts of speaking*" (e.g., Levy, Duncan, & Cassell, 2008, p. 6). The bodily performance that produces the mundane features of a lecture grounds the communication between presenter and audience. The observed heterogeneity (multimodality) of communication in a lecture is characteristic of human interactions generally. In what follows we highlight and discuss the significance of the results in the previous analyses. We articulate scholarly insights into the knowledgeability and improvement of teaching and learning concepts in lectures. We suggest that the heterogeneity of bodily performances is not a redundancy but constitutes the very terrain of communication that makes the development of concepts possible – i.e., teaching and learning as *hybridization*.

Teaching in Lectures as Hybridization

The presence of artifacts and their multiply layered relations to the human body constitute a distinctive aspect of scientific communication in lectures. Inscriptions (e.g., graphs) constitute prevalent resources that mediate talking concepts in physics lectures. From a perspective of the lecturing physics professor, his talk and physics inscriptions that he presents may all make clear sense and refer to a set of clear ideas (i.e., physics concepts). Yet, from a students' perspective, the use of inscriptions raises a (pedagogical) problem: Students depend on a professor's talk to know inscriptions presented at hand, but the professor's talk with/about those inscriptions is known to students only when they already know the inscriptions. In fact, we can hear – and therefore speak – only because we always already understand (Heidegger, 1927/1977), and, because understanding is associated with language, inscriptions *presuppose* but *are irreducible to* language. Coming to understand physics inscriptions and associated physics talk mutually presuppose one another.

Our analysis of the lecture performances shows that the body has a significant role in this pedagogical practice. Thus, the professor's iconic gestures of pushing and pulling in the second episode and associated utterances "compression" and "decompression" are constitutive for the sense of the associated physics concepts (adiabatic demagnetization and the third law of thermodynamics) by their analogous relation to the graph – they constitute a form of translation of non-identical expressions into one another. As translations go, they inherently capture only some but not other aspects of the original expression even in the special case where both expressions are within the same modality or language (Derrida, 1996). The lecturer's body orientation and eye gaze toward the graph realize this structural (cultural) relation and drive the analogy or metaphoric thinking. These heterogeneous body movements between spaces (i.e., between the graph and the narrative/gesture) not only open one additional space by differentiating the latter (narrative space) from the former (inscription space) but also explicate the conceptual presuppositions that make this translation possible. Therefore, we conclude that the professor's competence lies in the knowledgeable bodywork that translates between different communicative content. Teaching physics concepts means translating and hybridizing the ongoing lecture talk. The capacity to increase the heterogeneity of communicative resources (i.e., hybridization) constitutes the central aspect of knowing concepts. Physics instructors successfully assist their students to understand physics concepts by producing different communicative performances.

The role of the body in teaching physics concepts is related to the dialogic (social) nature of communicative practice. In lectures, the mobilization of cultural artifacts (e.g., pointing to a graph) presupposes the understanding of the other (i.e., audience). A physics professor may do this performance rather than something else because he is located in a specific condition of talking physics. For example, the mobilization of the compression-decompression analogy for adiabatic demagnetization makes sense on the assumption that students are familiar with the physical relation between volume-pressure-temperature while a gas undergoes an isothermal compression and an adiabatic decompression. The physical presence of students in the lecture hall provides a set of possibilities, some of which are concretely realized in the lecture performances. Listening to lectures is different from reading books alone because the lectures produce in heterogeneous communication a hybridized form of knowing that is not equivalent to any other. In this sense, the metaphor of the "construction of meaning," which is often used to theorize classroom communication, pertains only to the conscious parts of knowledge and not to all forms of knowing that make explicit forms of knowing possible in the first place.

Learning in and through Lectures: a Chain of Translation

The tenor of our chapter is that concepts are not ephemeral entities, metaphysically existing somewhere in the rarified spaces of another world, but constitute real, corporeal performances and living labor. This labor is observable in the performative productions of lectures. Physics concepts (heterogeneously) performed in lectures have the potential to be transferred and therefore translated into different sense-making units under the condition that the students perform similar work. Translation involves performances of mapping one domain onto another. Transformation means that dissimilar entities are made to correspond to one another, such as a line graph, on the one hand, and a mathematical equation, on the other (Janvier, 1987). Translation also occurs within the same communicative mode, such as when the word "dog" is mapped onto the words "Hund" (German) or "chien" (French). Translation occurs even within one and the same language (Ricœur, 2004). Such translation is commonly observed when a listener requests a speaker what he means leading to the speaker to use different words "to say the same." That is, translation requires an *assumption* that two different expressions are the same although they are very different on the surface. To what extent this assumption is fulfilled is an issue that interaction participants tend to solve in a pragmatic manner and as a matter of course. The process of learning physics concepts in and through lectures therefore can be seen as a chain of translations that involve the migration of lecture performances from the initial forms in which they are produced (e.g., university lectures) into those which students/professors produce (e.g., lecture notes) and also into other (sub-) cultural settings (e.g., individual study at home, and problem-solving with peers). Our examples show that some performances of physics concepts not only comprise verbalizations but also gestures, body orientations, and gazes.

Students, however, only record in their notebooks what is written on the chalkboard and some of what the lecturer says (Figures 2.4a, b). Students' notes generally lack all the other communicative resources that a lecturer makes available in modes other than talk, including gestures, body orientation, body position, and prosody. There exists a tension between the physics concept as it presents itself, performed by the professor, on the one hand, and the reduced way in which the lecture is recorded in the students' notebooks, on the other hand, where all the resources that the professor's body has made available in the talk no longer exist. We

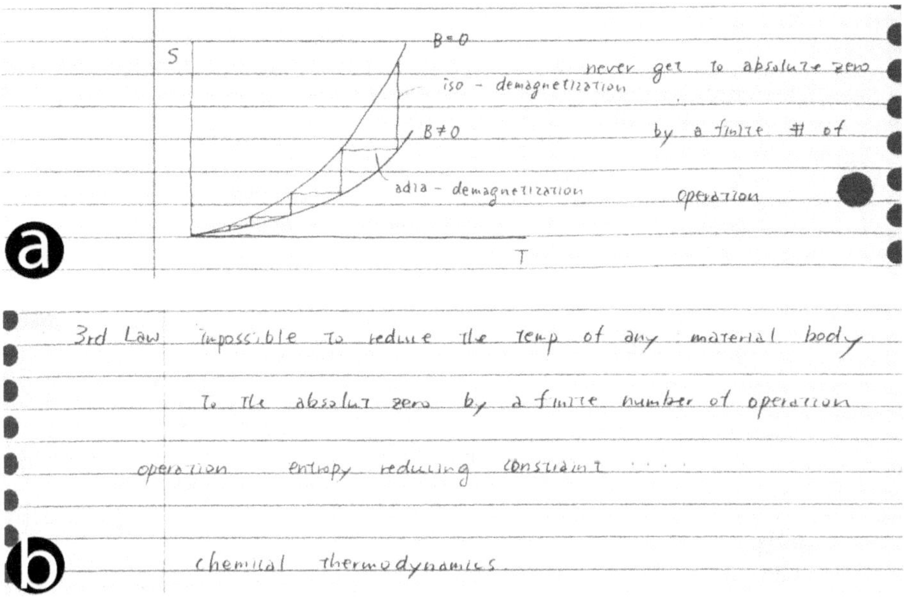

Figure 2.4. An example of students' class notes: it consists of a copy of the graph and the professor's verbal explanations (a) and some textual descriptions of the Third law of thermodynamics as stated by the professor (b). These notes do not comprise other communicative resources that the professor made available in the talk (gestures, body positions and orientations, and prosodies) and therefore capture only some aspects of the physics concepts performed in the lecture.

(the authors) are not astonished to find that students often do not make sense of their lecture notes, which capture only some aspects of lecture performances.

Lecture notes – and, for the same reason textbooks – do not replace lectures (Figures 2.5a, b). This, too, may explain why experienced professors often ask their students to look at them talking in lectures instead of being occupied with making notes or reading textbooks. Gazing at and listening to the professor places students in a strange position: They are familiar with the world and yet contribute to producing growth points in the lecture performance. Therefore, we suggest that students take different communicative performances as their resources for understanding concepts and also attempt to increase the possibility to mobilize them for their individual studies. Moreover, teachers and educators need to recognize that transferring some communicative productions to another setting brings about the reduction of the whole unit of information that was available in the previous setting and at the same time the amplification in the course of enacting those productions as part of another communication unit. Learning physics concepts means to become able to navigate the world in communication rather than "constructing" something independent of the concrete condition (e.g., abstract disembodied metaphysical concepts).

16.2 *The Third law of Thermodynamics*

Nernst's heat theorem and Planck's extension of it, while originally derived from observing the behaviour of chemical reactions in solids and liquids, is now believed to apply quite generally to any processes, and, in view of that, it is time to reconsider our description of adiabatic demagnetization. We see immediately that figure XV.1 needs to be redrawn to reflect the fact that the entropy of the substance approaches zero whether or not it is situated in a magnetic field. The revised drawing is shown in figure XVI.1, in which I have drawn three consecutive magnetization-demagnetization operations, and t will be readily seen that we shall never reach a temperature of exactly zero in a finite number of operations.

The same applies to any operation in which we attempt to lower the temperature by a series of isothermal constraints that decrease the entropy followed by adiabatic relaxations – whether we are compressing a gas isothermally and then decompressing it adiabatically, or stretching a rubber band isothermally and loosening it adiabatically. In all cases, owing to the convergence of the two entropy curves at zero temperature, we are led to conclude:

It is impossible to reduce the temperature of a material body to the absolute zero of temperature in a finite number of operations.

This is the *Third Law of Thermodynamics*, and is an inevitable consequence of Planck's extension of Nernst's heat theorem.

a

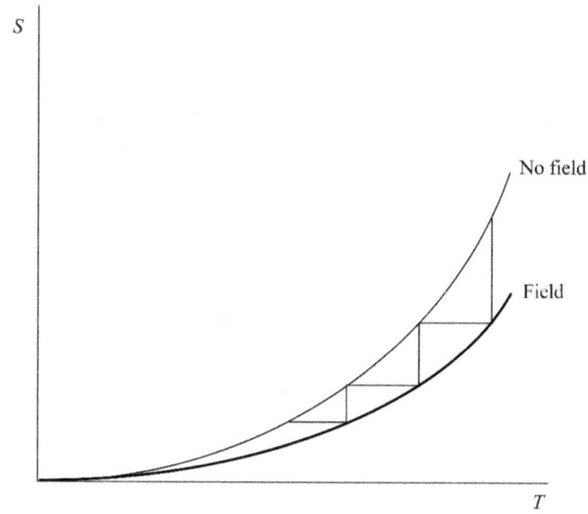

b

FIGURE XVI.1

Figure 2.5. Lecture notes that the professor handed out to students provide textual information of the Third law of thermodynamics driven from Nernst's heat theorem (a) and a graph that refers to magnetization-demagnetization operations (b). These lecture notes do not replace the lecture because they do not comprise all the performances that are made available in the real talk (e.g., the compression-decompression analogy).

FROM CORPOREAL PERFORMANCES TO PARTICIPATIVE (UNINDIFFERENT) UNDERSTANDING

In this chapter we articulate and exemplify a different way of understanding physics lectures. We specifically show how physics lectures constitute embodied performances, that is, work producing concepts. This work exists in and as of the lecturer's bodily engagement with things and in itself. We propose that the different communicative productions in and across different physical spaces of the lecture hall constitute the sense of physics concepts because the body renders them structured in the movement. Yet, the structure of those communicative performances does not pre-determine lectures because the movement of the body takes place in real-time within a situation as a whole. Thus, in our examples, the professor does not mechanically reproduce a planned talk but actually articulates concepts in real time for, and in the presence of, the audience. Inscriptions, such as graphs, constitute resources toward which the body orients itself; therefore they become an agent that participates in the ongoing articulation of physics concepts – some science studies theorize the role of materials and inscriptions in scientific work from the perspective of agency – the powers that reside in the sense of the phenomenological "I can" – and performativity. Inscriptions, textbooks, and words are constitutive parts of higher order communicative units rather than re-presenting concepts in themselves. Therefore, in science lectures, concepts are *heterogeneous* because they are distributed over the physical arrangements of different communicative performances that the body produces in relation to and with other things in the setting. The dynamic transactions between different modes of communication allow those resources to constitute a participative (unindifferent) understanding unit.

Given that there are many good physics books on the market and an increasing number of Internet resources, it is a legitimate question why universities offer lectures as part of their curriculum. What is it that lectures offer over and above verbal content and visual representations? Whereas quite a number of science education studies propose alternative forms of teaching other than lectures and report their effectiveness, it is rare to find communicatively oriented studies that explicate what makes lectures successful or unsuccessful. In this chapter, we provide an account of the value-addition when students attend lectures as compared to reading a textbook or an online resource, and thereby informs physics educators of the probable directions toward which research on physics lectures could be configured and evaluated – for example, the spatial arrangement and coordination of communicative fields have the power to explain the sense-making capacity of the resources provided in lecture talk other than the main, conceptual storyline. Our perspective also provides an explanation for the gap between understanding that occurs in a physics lecture and that which occurs during preparing or taking an examination. In the performances of physics professors, there is much more than words and images that assist students to make sense of physics concepts. Because of the many resources made available, lectures offer opportunities to *participative (unindifferent) thinking* (Bakhtin, 1993) on the part of the students. In this chapter we provide a first proposal for the "more" that lecturers make available over other forms of presentations such as textbooks or lecture notes. In fact, in a strong sense, any concept only exists in and as of its performance (Husserl, 1939); that is, concepts do not exist independently of the work that concretely realizes them in communicative acts. Just like a language that nobody speaks, mathematical and scientific inscriptions constitute only virtual possibilities of cultural knowledge, which always requires a performance to be kept alive. Books only provide resources but not the subjective work that actually constitutes the objectivity of the sciences. It is only because they are concrete-in-use that audiences can perceive concepts. By opening up, the different senses (eyes, ears) of listeners are affected; and this opening up to being affected through the senses is a basic requirement for the constitution of sense. The development of educational technology adopted in science (physics) lectures and talks increases the heterogeneity of communicative productions and therefore highlights the significance of cultural approach to communication adopted in this chapter. We now need to research ways in which students – and lecturers – can be assisted to be more aware of these additional sense making resources so that they can really capitalize on the value-addition that the lectures they attend offer to them.

PART B

THE ROLE OF THE BODY IN SENSE MAKING

> My perception is therefore not a sum of visual, tactile, and audible data: I perceive in an unreduced manner with my whole being, I grasp a unique structure of the thing, a unique manner of existing that speaks to all my senses at once. . . . Perception is not a sort of beginning science, an elementary exercise of the intelligence; we must rediscover a commerce with the world and a presence to the world which is much older than intelligence. (Merleau-Ponty, 1996, pp. 63, 66)

Understanding is founded on perception and perception on understanding. But perception is not the sum of individual sense data. Rather, all sense-making resources are perceived by the individual in and through her existence as a whole, undivided and indivisible being. All perceptions always already are *mine*, those of a single being in flesh and blood who does not have to construct correspondences across the different modalities of perceived communication modalities. All the senses are part of the same being, who integrates its experiences into the sense. In part A of this book, we articulate the double nature of sense experience in learning science and mathematics in which the living body takes the central role for the accomplishment of higher-order *thinking*. This dialectical framework of the development of participative (unindifferent) understanding provides an extended answer to the question of how a person comes to see the world (object) differently than before while engaged in *everyday* practice of doing (talking). One part of communication is conceptualized in terms of *verbal thinking* (e.g., L. S. Vygotsky). A more holistic stance makes it *participative (unindifferent) thinking* (e.g., M. M. Bakhtin, 1993). Thus, when people verbally think in the presence of scientific or mathematical artifacts, they express ideas in and through various forms of bodily action of which the sense is always co-determined by anything and anybody else present. Here, the central phenomenon to this (verbal, participative) thinking is that the living body performs a world (object) and in the course situates itself as a body among many bodies. The living body has been marked by very different experiences, which, operating together, constitute the sense that is common between the speaker and the hearer (e.g., student and teacher). That is, co-present bodies produce irreducible relations to the living body and come to play a significant role in marking scientific and mathematical sense. As a result, the participants come to see something *as* something. For example, scientists are trained with computer programs and for them computer interfaces are like organic parts of their individual bodies that act toward the world of which they are integral parts, feel the world, and are affected by it. In the same way, we may propose that students come to understand concepts because their bodies are capable of performing thinking in the presence of other bodies and are capable of being affected (hybridized) by the experience arising from their actions and the effects thereof on others. This reciprocal relation between the living body and other sense-provoking bodies in communication makes the living body a host that keeps altering sense experiences and thereby leads to see the world differently than before.

In this second part of the book, we further articulate and elaborate the role of the living body in sense making. We study the heterogeneous nature of sense-making praxis and how the living body enables and mediates it. We review major concepts of science and mathematics education from the perspective of the living body. We expand our previous discussion to articulate and answer some key problem of learning – e.g., the role of text, artifacts, visual representations, and analogy in classroom communication. Our aim is to thematize the mundanity of learning science or mathematics through common, everyday classroom praxis. That is, we develop a theory of sense making centrally mediated by the living body rather than assuming any form of representation that stands on its own, apart from the living person. In philosophy, the role of the living body in making sense is acknowledged (Nancy, 1993b). On the one hand, the living body is the source of singularity – there is only one living body in a particular place with particular experiences. Because it takes up a different spatiotemporal position in communication from others living bodies, it becomes a referent that is marked with categories (boundaries) such as culture, gender, and race. Yet, on the other hand, the living body is the source of plurality (alterity) because in its orientation toward the other (e.g., speaking, listening) the space that it occupies and moves across is already inhabited by the (generalized) intelligible Other who perceives my body as singularity from her own singular perspective – that is, my own body is intelligible on the presupposition that it is already a body among bodies (Ricœur, 1992). In this intersubjective space the act toward the Other simultaneously touches (affects) the self – the living body is affected and marked by the act it has performed and from which it learns about its own thoughts. Communication expands existing possibilities of knowing and leads participants to see what they have not yet known. Therefore, we understand that Vygotsky's concept of the *zone of proximal development* involves this nature of human communicative action – both individual and cultural – and the heterogeneity (hybridity) of communication in learning rather than it constituting an artificial platform directed toward only an individual child in separation from an adult (i.e., scaffolding).

In the following four chapters, we articulate and elaborate the role of the living body in sense making as the interface of mind and culture – the interface here means that sense making is neither the matter of an individual mind nor a culture but the event in its entirety. In fact, it is difficult to locate an interface, though in our personal experience we make distinctions between Self and the Other. But the consciousness we have of the self, self-consciousness, inherently means knowing (Lat. *sciēre*) with others (Lat. *con-*) about the Self. This relation of Self and all others can be understood in terms of the same analogy that we provide for concepts (Figure 2.3). Here, any individual "me" is a function of the relation to all other selves. Anything I know and anything I am is other, has come from the other and, in my speaking, returns to the Other. Each chapter attends to the lived experience of sense making and the encounter of human intellectual functions (i.e., thinking) and communication comprising cultural tools. The integration of psychological and sociological themes is achieved by highlighting the role of the living body that imparts sensuous experiences and also the experience of their whole. In chapter 3 we propose theorizing literacy as the performance of the living body rather than the mental or the purely linguistic capacity. By using an example of doing an experiment in a university physics laboratory, we show that the body is deeply involved in knowing a language, which is undecidably intertwined with knowing one's way around the world consisting of physical tools, equipment, and micro computers in this case. In chapter 4 we expand the relation between communication and the experience of the world and therefore further theorize the role of the body in conceptual development. In chapter 5 we theorize the learning paradox involved in knowing mathematical inscriptions and we study the role of the body in making something stand for something else. In chapter 6 we theorize the relation between everyday language and science language and discuss implications for pedagogical discourse and scientific and mathematical concept development in the classroom.

CHAPTER 3

LITERACY AS BODILY PERFORMANCE

> Language is not in the world or inside the world, as though the world were its body: it is the outside of the world in the world. It is the whole of the outside of the world; it is not the eruption of an Other, which would clear away or sublimate the world, which would transcribe it into something else; instead, it is the exposition of the world-of-bodies as such, that is, as originarily singular plural. (Nancy, 2000, p. 84)

Literacy denotes the cultural nature of knowing to communicate. In learning science and mathematics, communication that comprises words, visual images, mathematical representations, and technological equipment constitutes the important part of enculturation into scientific and mathematical culture. Previous studies of literacy often use the notion of "constructing meaning (sense)" to theorize the process of communication by means of which students get in contact with scientific and mathematical bodies and become literate. Yet, the conditions that make literacy practice (i.e., mediation by tools/signs) possible and bring about the associated development of higher cognitive functions are hardly theorized. Therefore, literacy is often attributed to psychological categories independent of the bodily practice (e.g., reading as mental construction). In this chapter, we develop the idea that literacy always expresses itself in the (discursive) actions that are corporeally produced and performed. That is, we articulate the central role of the living body in the performance of scientific and mathematical literacy. We conduct an exemplary analysis of literacy practice and the associated development of conceptual understanding in a microcomputer-based physics laboratory to support two claims. First, the body temporally engages in (laboratory) artifacts located at different places, and thereby spatially realizes literacy. Second, the body spatiotemporally coordinates cultural artifacts in the public terrain where higher-order cognitive functions first exist. We conclude that the central aspect of literacy in students' conceptual development consists in the engagement of the body in the world, which bridges sense-making resources and the understanding of the world that these metonymically denote.

THE PROBLEM OF LITERACY

Literacy is a core goal of science and mathematics educators. In classrooms concepts are made available to students through talk or written textual materials involving words, graphics, and mathematical representations. Literacy practice – such as listening to the others talking and reading representations – allows students to participate in classroom communication and grasping concepts. The problem with the concept of literacy as used in educational studies is that it does not address all the extra-linguistic capacities that are part of any practice (e.g., Schutz, 1996). There is agreement among language philosophers with a phenomenological bend that linguistic structures and everything else that makes life are woven together into an irreducible tissue such that knowing a language is equivalent to knowing one's way around the world. In this framework, literacy is defined in terms of participating in relevant everyday activities (e.g., Roth & Barton, 2004). That is, literacy involves competency in "social actions [that] presuppose communication and communication necessarily is grounded in working acts" (Schutz, 1996, p. 32), which require the living body as a whole. Thus, I am computer literate when I flexibly use computers to achieve my personal goals. I am scientifically literate when I mobilize scientific resources – concepts, theories, or scientist consultants – for achieving my purposes.

Some psychological studies take a constructivist approach and theorize literacy practice as "constructing meaning." Yet, the metaphor of construction pertains only to the conscious parts of knowledge and not to all those communicative means that make explicit forms of knowing possible in the first place (Roth, 2011b). For example, in a typical physics laboratory, students are provided with a manual and a lab monitor's instructions concerning the experiment, the method of using various kinds of equipment, and the procedure to collect and analyze data. In this situation, literacy implies appropriate use of textual materials, verbal talk, and laboratory equipment. Since the relevance of reading/writing or speaking/listening presupposes students' understanding of the laboratory work and physics concepts related to them, which students do not get into yet before doing the experiment, this raises questions: What is it like to become scientifically/mathematically literate and act relevantly? How do scientific/mathematical things (material bodies) appear to students and how do students experience them in literacy practice? The purpose of this chapter is to contribute to the development of a theory that reviews literacy through a perspective of the social world that "starts from the actor's subjective point of view" (Schutz, 1996, p. 9). We propose a theory of literacy that does not dichotomize the senses and sense both of which together constitute the real condition of the actor's subjective experience of and on the world and its expansion. We draw on an ethnographic study of knowing and learning physics in a computer-based laboratory to exemplify aspects of literacy development arising from the body engaging in the world.

A GENETIC AND HOLISTIC APPROACH TO LITERACY

Literacy is commonly studied by analyzing specific forms of representation (e.g., words, numbers). Pencil-and-paper type questionnaires are administered in various settings of science and mathematics education to evaluate students' understanding of concepts on the assumption that they appropriately re-present the corpus of knowledge to be studied. Sometimes, a broader concept of literacy is adopted to include the practical dimension of science and mathematics to the metaphysical definition of conceptual knowledge, but it still keeps the separation of linguistic representations from the bodily practice when, for example, the latter is categorized in terms of skills apart from the conceptual content knowledge. However, the separation of linguistic representation and its sense that is marked in and through communication is neither fruitful nor possible (Bakhtine [Volochinov], 1977). From a dynamic perspective of development, the significance of literacy in learning science and mathematics lies in the provision of possibilities to expand the actor's power to move onto a realm of the not-yet-known. That is, the concept of (scientific, mathematical) literacy needs to indicate not only the power to competently engage in tasks that a person is already familiar with but also the empowerment to find one's way in the new and unfamiliar world (e.g., physics laboratory). Literacy then is something like fluency, the fluency of moving about in a world without having to make it or the movement thematic. Fluently moving about is like walking, which we do without cogitating how to do it or where to place the feet next.

In this chapter, we take a dialectic approach to thought and language each of which is one-sided to a larger unit that encompasses the fullness of life (Roth & van Eijck, 2010). We thereby attend to the possibility of explicitly theorizing the dynamic relation of knowing a language and knowing one's way around the world – i.e., the role of literacy in sense making in scientific and mathematical communication. We propose that a holistic theory of literacy has to include the dynamic role of the living body, which allows, for example, emotion to become an integral part of thinking in the way Vygotsky asked for. Literacy then may express itself concretely as part of a corporeal life that is irreducibly interconnected with language. The following four points, which reiterate theoretical claims in the first chapter of this book, summarize and guide the genetic, holistic approach to literacy. First, the body constitutes a hub for an undivided experience of the world (object). Second, the body constitutes a hub of communicating scientific and mathematical (written or spoken) text. Third, communicative action in and about science and mathematics in the presence of scientific and mathematical artifacts (e.g., machine) constitutes a unit that is produced by a corporeal "I can" that integrates the different communicative modalities (e.g., speech, gestures, eye gaze, body orientation and movement). Forth, eye gaze, gestures, body orientation and movement, which are involved in experiencing scientific artifacts, are also involved in communicating science and mathematics. To sum up, literacy is observable in the form of performances of the living body that produces different communicative forms. These forms may have different, even contradictory content and yet constitute one whole.

Human-Computer Interaction and Sense making

Scientific and mathematical communication comprises various cultural (disciplinary) means. The recent development of technology has made computer-assisted apparatuses indispensable to learning. For example, students learn about the properties of physical phenomena (e.g., heat, light, etc.) through computer systems equipped with sensors and programs that produce mathematical representations (e.g., graphs) with measurements. In an undergraduate experiment on a thermodynamic cycle, students may use computer sensors attached to a gas cylinder to measure instantaneous values of gas volume (*V*) and pressure (*P*) and also work with data-processing computer-software to transform obtained measurements into a *P-V* diagram. Therefore, previous studies in education showed the affordances and constraints of utilizing computers in teaching and learning science (Roth, Woszczyna, & Smith, 1996). Computers provide both students and teachers with opportunities to spend more time talking about physical phenomena and data collected rather than on the routine operations for drawing tables and graphs. However, the use of computer interfaces raises a new problem; students have to spend quite some time on developing their senses and sense about new tools and mediated phenomena. Despite various recommendations for effective computer-supported science classroom, one fundamental issue here is that human beings are actively and physically involved in making computers work appropriately in a given situation. That is, understanding the possibilities that the computer technology provides for sense making, which is of significance to literacy, requires studies on human-computer interaction and the process of becoming competent in the situation of using computers.

In the case of a physics laboratory, computers may constitute an interface between the student and some physical phenomenon, which is the object of the student's inquiry. Computers are constitutive moments of the relationship between the human subject and the phenomenal world. On the one hand, the computer interface constitutes part of object; students' knowledgeability of the thermodynamic cycle, for example, is closely related to their learning to use the computer. On the other hand, as students become competent to looking at phenomena through instruments, the computer is an extension of the person and becomes part of the acting subject. Students make sense of the phenomenon as if the computer interfaces were part of their bodies (eyes, ears, or hands). Therefore, human and machine do not make mechanical input-output relations but constitute inseparable unit in communication. The role of the computer as part of the dialectical unit of subject and object constitutes a central aspect of understanding human-computer interaction and thereby, the development of literacy in the science laboratory. Dialectical unit here means that we know the nature of the subject of action and object only from the subject–object *relation.* When the relation changes, then the subject of action changes *even if it is the same individual.* This is so because the activity as a whole determines the nature of subject and object (Leont'ev, 1978).

The dialectical relation between subject and object can be conceptualized by attending to the role of the living body, which, as Marx/Engels (1962) point out, does the "living labor." The living body engages with a computer as an extension of both experimental equipment and the subject, who senses (with) the equipment. Vygotsky thinks of such a relation as one of mediation. The living body interacts with the computer interface and unfolds a series of action from an initial step of calibrating for the instrumentation for collecting data to a final stage of producing various forms of inscription (graphs, numbers, tables). In this sense, living bodies lie at the center of understanding students' knowing and learning in computer-assisted laboratory. It is through their sensuous labor, produced in, by, and with living bodies that the equipment produces anything of interest. In the following we take a closer look at one case to exemplify the living labor of sense making and the associated development of physics concepts. We show how a student engages in the world of an undergraduate thermodynamics laboratory and how she thereby develops an understanding of the topic (i.e., Stirling cycle). We substantiate two claims. First, in/through her living body she engages with different laboratory artifacts and realizes literacy in a spatial form. Second, the body spatiotemporally coordinates cultural artifacts, which constitutes a terrain from which a higher-order cognitive function arises.

MAKING A COMPUTER INTERFACE WORK: LITERACY AS LIVING LABOR

The concern of literacy in this chapter is related to the (temporal) development of sense rather than "meaning" that is assumed to reside in words. In fact, pragmatic philosophers exhort us to abandon the use of "meaning" because it is a concept that derives from a primitive conception of language (Wittgenstein, 1953/1997). This very concept is part of the ideology that splits the body from knowing. Because our living bodies constitute the position that we take in and with respect to the world, we focus on the living body, its senses, and on the sense it makes. We do not only experience our bodies in absolute spatial, measurable time; rather, we also experience "events in inner time (*durée*) . . . as manifestations of our spontaneity" (Schutz, 1996, p. 29). Listening to a talk or reading a text not only unfolds in and through time but also makes time. Physics undergraduate students participate in conducting an experiment in a laboratory and temporally develop their understanding of topics. Time is generated, as the pedagogical possibilities are made available through talk and working acts with laboratory resources. Certain ways of being emerge, which we also observe in the following case example culled from the beginning phase of a three-hour thermodynamics laboratory experiment. "Stirling engine" is the topic of the experiment and the apparatus consists of a piston-type engine, correlated with a computer interfaces, and other measuring equipment such as a power supply, a voltmeter, and an ammeter (Figure 3.1a, b). In the following, we do not observe disembodied minds do something but living labor that arises from the engagement of real, living bodies in the world (objects). The undergraduate thermodynamics laboratory constitutes a space in which students' working acts (e.g., making interfaces work properly, talking) constitutes the praxis of literacy (i.e., communication about Stirling cycle).

The lab instructor Carl stands near Mariko's bench and gives an introduction of the experiment to Mariko and another student (Andy). Carl is talking about how to calibrate the initial conditions of the Stirling engine (maximum, minimum values) connected with the computer interfaces and get the computer started to collect empirical data from a pressure sensor and a volume sensor (Figure 3.1). As the wheel of the engine turns, the top and bottom plates of the working piston move up and down and the volume between the two reaches a maximum and a minimum in turn. As the working piston inside the cylinder undergoes a cyclic change of the internal

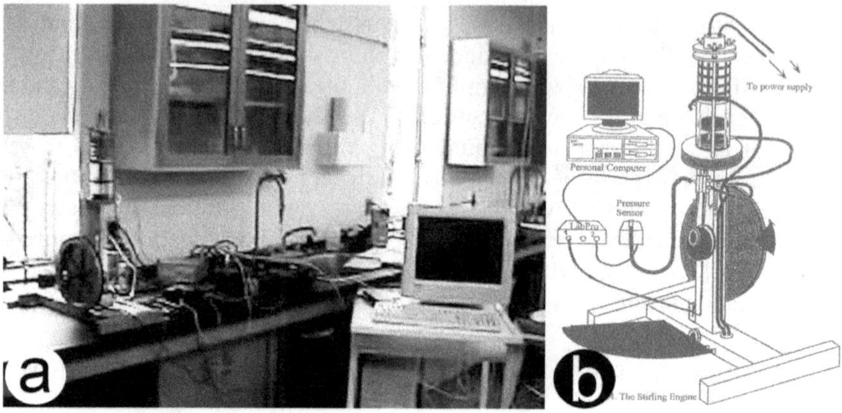

To set up and calibrate the LabPro for data collection, first turn on the d.c. power supply hooked up to the slide-wire potentiometer and set the voltage to a value slightly less than 10 V. Next, click on the LoggerPro icon on the PC desktop and select for a new experiment. Click on *Setup* → *Data Collection* → *Sampling* and enter **250 ms** for the *Experiment Length* and **.1 samples/ms** for the *Sampling Speed*. Again, under *Setup*, go to *Details* and click on the *Unlock* button. Click on *CH 1* and enter the label as **Volume** and units and short label as **cm^3**. Next, go to *Calibrate* and perform a two-point volume calibration by clicking on *Perform Now*. For Reading 1, enter **310** cm^3, maximize the voltage reading by rotating the flywheel of the Stirling Engine, and accept the maximum voltage value when it is obtained. For Reading 2, enter **155** cm^3 and obtain the minimum voltage in a similar manner as before. Now, return to *Sensor Setup* and select *CH 2*. For the sensor, select **Pressure-Pressure Sensor**. Return to *Details* and lock the sensor calibrations. Both channels are now calibrated, and the engine may be started.

c

Figure 3.1. a. The Stirling Engine set up in the laboratory. b. The diagram of the Stirling Engine presented in the lab manual. c. The calibration method written in the lab manual.

volume, the gas inside experiences compression and decompression. The computer software (LoggerPro®) records data from the sensors attached to the engine (DataPro®) and produce mathematical representations of thermodynamic cycles (i.e., graph).

In Episode 3.1, Carl suggests capturing a minimum of five cycles to be able to empirically determine the pressure-volume relation (turn 01). Andy, who stands on the other side of Carl, reads aloud a part of the lab manual, "Again under *setup* go to *details*" (Figure 3.1c), and points out that the menu option does not exit in the computer program (turn 02). Carl suggests "go to setup and sensor in details" (turn 03). Mariko reads aloud her copy of the lab manual with very low speech intensity, as if reading to and for herself (turn 04). Carl asks whether they have "any questions" (turn 05), and thereby gives the two students an opportunity to clarify their questions. One by one both of them utter "not yet" (turns 06, 08). Carl moves away from them to assist other students. Andy goes back to his lab bench located at the other end of the classroom. The two students begin to work individually.

Episode 3.1

01 Carl: *then let it reach like equilibrium then you're ready to start taking data, okay ((*Mariko nods*)) and data taking goes very fast, when you're at that point we'll look at it and talk about it, at that point what you need to do, you're gonna basically, you're gonna just capture like five, five cycles, a minimum of five cycles, and each cycle is only a quarter of a second, it doesn't take very long to uh to get your data [Figure 3.2a]
02 Andy: and here it says again under setup go to details um that doesn't exist
03 Carl: oh, go to set up then sensor then details, okay?
04 ((*Mariko reads aloud some text of the lab manual that she holds in her hands. The contents of her speech is inaudible.*))
05 Carl: any questions about this at all?
06 Andy: not yet
07 Carl: okay ((*Lab instructor turns his body toward Mariko.*))
08 Mariko: not yet
09 Carl: okay ((*Lab instructor turns on the power supply and moves away.*))

Episode 3.2 (Continuing from Episode 3.1)

The following constitutes an ethnographic description of the events that can be gleaned from the video. Relevant video images are reproduced in Figure 3.2. The time on the video is indicated in round parentheses.

(00:00) *Mariko sits at the computer table and gazes at the lab manual. She turns on the computer (Figure 3.2b). (00:17) Mariko logs in the computer and continues reading the lab manual. (01:21) The camera angle changes to the engine on the lab bench. Mariko is not visible. (01:39) Andy stands besides Mariko's computer. She is talking to him. (01:43) Andy walks away and Mariko gazes at the lab manual on the table. (01:53) Mariko gazes at the computer screen (LoggerPro window) and opens different menu bars. (02:37) Mariko talks and Andy comes to Mariko's computer table. He opens a window from a menu bar. (02:55) Mariko keys in some information in a blank. A red warning message appears at the bottom of the window. (03:03) Andy talks and keys in some information. He moves away. (03:20) Mariko gazes at the lab manual. Her gaze moves between the lab manual and the computer screen. (04:05) Mariko puts the lab manual down on the table and keys in some information in the window while gazing at the lab manual.

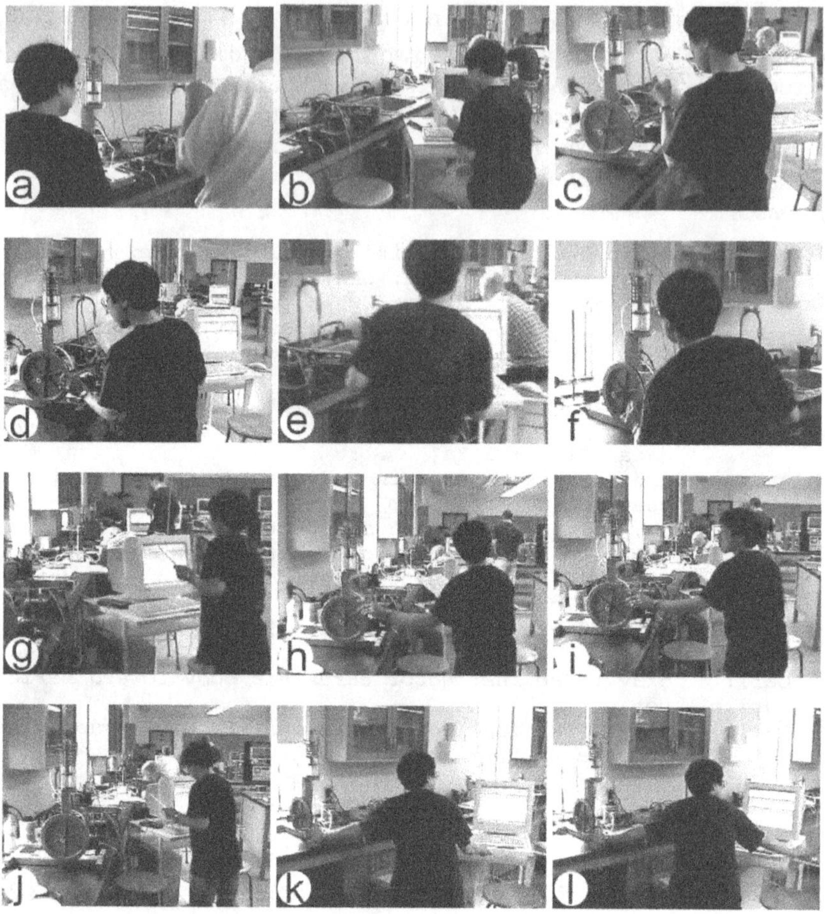

Figure 3.2. a. Carl (the lab instructor) stands in front of the Stirling engine and talks about the experiment. Mariko stands on his left side and Andy on his right side (invisible). b. Mariko sits at the computer table and reads the lab manual. She turns on the computer. c. Mariko stands up while reading the manual. She moves to the Stirling engine. d. Mariko rotates the wheel and gazes at the piston around the maximum volume. e. Mariko moves to the computer and gazes at the screen. f. Mariko moves to the Stirling engine and gazes at the inside of the engine. g. Mariko moves to the computer and gazes at the screen and the lab manual. h. Mariko moves to the engine, holds the wheel, and turns her gaze toward the computer screen. i. Mariko turns her gaze back to the engine. j. Mariko moves to the computer and keys in some information. k. Mariko holds the wheel with her left hand and gazes at the computer screen. She stretches her right arm and pulls the computer table toward her body. l. She gazes at the computer screen and grabs the computer mouse on the table.

(04:51) Mariko stands up. She glances at Andy's side. Andy is not at his table. She stands still and gazes at the manual. (05:08) Mariko slowly turns to the left and moves to the Stirling engine while gazing at the lab manual. *She stands in front of the engine and gazes at the manual (Figure 3.2c). (05:18) Mariko gazes at the engine and stretches her left hand to the wheel of the machine. She slowly rotates the wheel. (05:26) Mariko gazes at Andy's side. *She turns the wheel and gazes at the two-point piston moving. The piston reaches the maximum volume and returns (Figure 3.2d). (05:42) Mariko gazes at the lab manual and moves to the computer. *She gazes at the screen (Figure 3.2e). (05:51) Mariko walks sideways and stretches her left hand toward the wheel of the Stirling engine. *She slightly turns the wheel and gazes at the working piston moving at around the maximum volume (Figure 3.2f). Mariko moves to the computer and *gazes at the computer screen (Figure 3.2g). (06:17) Mariko moves to the engine, holds the wheel, and *turns her gaze toward the computer screen (Figure 3.2h). She slowly turns the wheel of the machine while gazing at the computer screen. (06:26) *Mariko changes her gaze at the engine (Figure 3.2i) and then to the computer screen again. She turns the wheel of the engine to the maximum volume, lowers down and the up again. Mariko gazes at the screen and glances at the piston as she makes the maximum, a decrease, and an increase of the volume (06:51) Mariko takes her left hand away from the wheel and tilts her body toward the computer. She gazes at the screen and tilts her body to the left toward the engine. She grabs the wheel and while gazing at the screen. She quickly leaves the wheel and tilts to the computer. She turns to the engine and holds the wheel. She gazes at the piston and turns the wheel. She adjusts the position of the piston to make the maximum volume. (07:22) Mariko glances at the lab manual and moves to the computer. She gazes at the computer screen and gazes at the lab manual. She moves to the other apparatuses placed beside the engine and gazes at them. (07:35) Mariko walks to the computer. She gazes at the lab manual. She quickly points at a part of computer screen and *keys in some information (Figure 3.2j). (08:17) Mariko stretches her left arm and holds the wheel. She gazes at the computer screen and turns the wheel. She glances at the position of the working piston and in turn gazes at the computer screen. She reads the lab manual on her right hand and looks at the screen. (08:56) She moves the lab manual from her right hand to the left hand. She stretches her right arm to the computer mouse. She tries to grab the wheel with her left hand that actually holds her copy of the lab manual. She gazes at the machine and put the paper on top of a chair. She holds the wheel and gazes at the computer screen. *She pulls the computer table toward her body (Figure 3.2k) and *grabs the computer mouse on the computer table (Figure 3.2l). (09:16)

In this situation, Mariko listens to the lab instructor's introduction to the topic and equipment (Episode 3.1) and conducts calibrations of computer interfaces (Episode 3.2) following the instructions specified in the lab manual (Figure 3.1c).

The text describes the procedure for inserting the amount of the gas volume following the position of the two-point working piston meeting each of the maximum and minimum volume conditions (310 cm^3 and 155 cm^3, respectively) – the gap between the two (top and the bottom) plates of the piston constitutes the volume enclosed inside the cylindrical chamber (Figures 3.1a, b). Mariko holds her own copy of the lab manual and reads the instruction all the while she works on the computer and the engine to make the computer interfaces work properly and to record right values of the thermodynamic cycle. Mariko's frequent orientation (e.g., eye gaze) toward the text thereby indicates that the text in the lab manual constitutes one of the resources (together with the computer interfaces) for scientific literacy that she has to express with respect to the physics topic. Her concrete working action with and on the equipment is related to the reading of the text: equipment and language come together in the same space.

Scientific literacy in this laboratory is not simply a matter of words but deeply involves individuals' engagement in the world, which requires the living work, for example, the continuous change of body position and body orientation. In this physics laboratory, Mariko has just begun to work and does not yet know what she would ultimately be doing with those scientific artifacts. She encounters something "which provokes sense without being meaningful itself, but still something *by which* we are touched, affected, stimulated, surprised" (Waldenfels, 2004, p. 238). Thus, the series of action in the Episode 3.2 (Figures 3.2b–l) shows that Mariko is engaged in letting "something" appear and grow in and through her active engagement in the spatially distributed sense making resources (e.g., verbal talk, textual instruction, computer, engine). The body position and orientation toward the lab equipment constitute resources that allow Mariko a specific way of being in the world and a specific way of knowing the world. This assumes and produces time and temporality (e.g., iteration). The living body here constitutes a hub that mediates between the words in the textual world and the world that she presently experiences. Although the words are also there, printed on the page, they also are something that is different in that they relate to all other forms of text (Ricœur, 1991). That is, the world that Mariko inhabits is experienced directly, whereas the words (sometimes) seem to come from elsewhere, some foreign/strange location, and therefore may not always make sense. Reading is living labor; and it is thereby, through Mariko's lived work, that the foreign comes to be related to the directly experienceable and experienced. The following two statements summarize the role of the body to the literacy practice in the physics laboratory.

First, the living body temporally engages with different laboratory artifacts and realizes (scientific) literacy in the form of spatially distributed performances. In the above example, Mariko's body movement appears in a space that consists of the computer table on the one hand and the Stirling engine (on the lab bench) on the other hand. The living body occupies and dwells in this space and works to make sense of things around it. Initially, Mariko sits at the table and operates the computer software while getting help from her classmate (lines 05–07, Figure 3.2b). She spends time reading the computer screen and the instruction by turns. Then, she stands up and walks to the engine on her left side (lines 18–21). She reads the same instruction in front of the engine (lines 21–22, Figure 3.2c) and same instruction in front of the engine (lines 21–22, Figure 3.2c) and turns the wheel carefully while gazing at the two-point piston working inside the cylinder (lines 23–27, Figure 3.2d). She turns to the computer and gazes at the screen (lines 28–29, Figure 3.2e). She turns back to the engine and gazes at the up-down movement of two working pistons while turning the wheel (lines 29–33, Figure 3.2f). She moves to the computer (lines 33–34, Figure 3.2g) and back to the engine again (lines 34–35, Figure 3.2h). The series of bodily performances concretizes a form of fluency – i.e., literacy in the space of the laboratory – by realizing cultural possibilities that the lab equipment and the text provide. The living labor connects the two artifacts, and thereby realizes the conditions of Stirling cycle – the Stirling engine makes a cyclic motion within the limited range of volume and pressure values.

Second, the body coordinates cultural artifacts in time and space. This constitutes a terrain from which higher-order cognitive functions arise. In the next, Mariko's body movement across the space turns out to be coordinated as a specific spatial configuration. She holds the wheel turns it back and forth (lines 35–37). Rather than walking toward the computer, she only changes her gaze between the pistons and the measurement screen (lines 35–39, Figures 3.2g–3.2i). The bodily alignment of temporal sequence to a spatial configuration allows Mariko to see the relation between the piston movement and the measurements on the screen (i.e., cyclic processes between a maximum and a minimum [lines 39–43]) and thereby to realize the conceptual possibilities involved in this mediated access to physics phenomena – i.e., by keying in appropriate information (lines 55–56, Figure 3.2j). Moreover, this alignment constitutes a terrain from which an even aligned spatial configuration arises: Mariko puts the paper down on the table and pulls the computer table toward the engine (lines 23–27, Figure 3.2k). Her right hand controls the computer and her left hand turns the wheel (lines 67–68, Figure 3.2l). The more structured spatial coordination by the body realizes the computer interfaces as an interface between the student and the phenomenon (Stirling cycle) and therefore increases *room to maneuver* (conceptual possibilities) for calibrating the interfaces in a way that measures the cyclic movement of the Stirling cycle – the performance of conceptual understanding.

This example from the undergraduate physics laboratory shows that the development of higher-order cognitive functions (i.e., calibration of interfaces) takes place not in a metaphysical (mental) space but in a space that involves sense making resources and spatial orientation made by the living body from one to another. The living labor produces literacy and extends the computer interfaces – they become part of the object of study (i.e., thermodynamic cycle of Stirling engine) and of the subject, who makes sense of the thermodynamic engine in the computer-assisted physics laboratory. Therefore, becoming fluent in using computer interfaces comes down to the issue of the spatiotemporal coordination of the living body with the interfacing artifacts, and thereby with the physical phenomenon for the study rather than some capacity solely belonging to an individual. Therefore, the analysis shows that researchers have to take account of the living body to understand fluent performance in the world, and therefore literacy. This means that they have to take into account the cultural nature of knowing and learning in this discursive space. Scientific literacy always expresses itself, concretely, in the (discursive) actions that are bodily produced and performed. The spatially distributed working act with respect to sense making resources makes the development of the associated higher cognitive functions possible. Human-computer interaction conceptualized in this way enables to see cultural nature of learning on the ground of the material world and the spatiality of sense making.

EMPOWERING SENSE MAKING

Knowing and learning to communicate scientific and mathematical topics constitutes a core goal of science and mathematics education since it empowers students to actively engage in cultural resources and capitalize on cultural possibilities. In this chapter, we exhibit where the empowering potential of literacy comes from: what is given to human beings engaged in grasping a physics topic that is not yet known to them? We exemplify that knowing and learning in the physics laboratory does not directly come from either verbal/textual instructions or scientific equipment. In our example, the student participates in a cultural activity (i.e., university physics laboratory) and simply works with the text and scientific equipment given to her. We understand becoming literate as the increasingly fluent performance in this for Mariko new world. The alignment of her working act in the discursive space emerges as she participates and encounters the (unfamiliar) words and scientific equipment rather than through the intentional act toward an outcome that already would have been known to her. Because the living body is open to the world and available to be affected because of its senses (Bourdieu, 1997), the totality of the field that the living body encounters in its mundane engagement with the material and social world of the laboratory leaves traces. In other words, the field comes to be inscribed in the living body. The very position of the subject in some disciplinary field shapes its dispositions, that is, its ways of perceiving the world and acting toward it. The very structures the living body encounters structure it, leading to a homology between field and dispositions, which Bourdieu (1997) calls *habitus*. With Foucault (1975) we might say that the material structures of the world impose a certain (physical) discipline, which characterizes the (conceptual) discipline of which a student becomes part.

The university laboratory is a cultural field in which a working act marks sense and is marked by the (pedagogical) dialogue that occurs over a period of time. This dialogic nature of living labor explains scientific and mathematical sense making without having to draw on psychological models that prescribe some mental construction to which we have no access. Pedagogically, the body-centered approach provides a different way of understanding scientific and mathematical literacy: as fluent engagement in the mundane world. Fluency means that the performance is adapted to the structural properties of the field, which includes, in the physics laboratory, the equipment, devices, language, and interactive forms. Literacy means that the student comes to navigate this world, knows her way around the world. Mariko's repeated reading of the lab manual is helpful as long as it lets her face the equipment in the laboratory and work with it. Some of the questions science educators may have include "What are the central aspects of the interaction with computers in knowing and learning science?" and "What makes the computer's mediation possible in the scientific understanding of phenomena?" The example in this chapter suggests that the computer mediates the relationship between human users and the phenomenal world that they are supposed to learn about in such a way that we cannot understand the actions, knowing, learning, or identity of the subject independent of the living body. The computer extends the object into the living body because it structures the way in which a student engages with the world. This engagement leaves traces: it in fact structures and thereby disciplines the body so that it comes to know its way around the world without having to represent it in the mind.

CHAPTER 4

THE BODY IN/OF MATHEMATICAL CONCEPTS

> Objects then become the transcendent ensemble that reveals my incarnation to me. A contact is a caress; that is, my perceptions not the *utilization* of the object and the surpassing of the present in view of an end; but to perceive an object, in a desiring attitude, is to caress myself with it. Thus I am sensitive, not so much to the form of the object and to its instrumentality, as to its matter (gritty, smooth, tepid, greasy, rough, etc.) and I discover in my desiring perception something like a *flesh* of objects. . . . From this point of view desire is not only the clogging of consciousness by its facticity, it is correlatively the ensnarement of a body by the world; and the world is made *ensnaring*; consciousness sinks into a body that sinks into the world. (Sartre, 1943, p. 432)

The classical conception of mind separates it from the things that it is conscious about. In the introductory quote to this chapter, Jean-Paul Sartre rejects such a division by stating that objects reveal my incarnation, which they can do only when their structures and those of my body are homologous. There is an interlacement in which my living body and the world come to be intertwined, mutually structuring each other. It is not only, therefore, that mathematical objects comes to shape the practices students develop but also the body comes to be engulfed in the world. In and through my flesh, I discover the flesh of objects, which, in turn, reveals my incarnation, that is, my corporeal nature that is thought by means of the category of the flesh. My body is in and indistinguishable from the body of mathematical concepts.

Pedagogical theories often discuss conceptual development in terms of the literacy of an individual child or teacher. These theories give little attention to the real conditions that make the mundane communication of an idea possible in the first place. The mundanity of everyday communication and its non-mentalist nature has been worked out in philosophy nearly a century ago (e.g., Heidegger, 1927/1977); and yet educators continue to draw on "ideas of a language more primitive than ours" (Wittgenstein, 1953/1997, p. 3). In this chapter, we extend our pragmatic approach to literacy and participative (unindifferent) understanding to give the body a central role in conceptual development. We exemplify the assertions that (a) the body, which produces living labor, constitutes something like a hub that translates, in an ongoing manner, the experience of the world (objects) and the word in communication; and (b) the living body constitutes something like a hub that generates new ways of making sense of the word. We conclude that a framework centered on the living body helps us understand the living and concretely lived labor that increases the corporeal resources for translating participative (unindifferent) understanding.

In many educational practices, pedagogical theories are usually about instructional strategies or instructional models based on (frequently unquestioned) presuppositions about the learning process. Those theories do not explain the real experience of teaching and learning, which is indicated by the real experiential gap between (pedagogical) theory and (teaching) practice (e.g., Roth, 2011c). One way of approaching this pedagogical issue is to rethink how teachers and educators theorize language and literacy. One study showed, for example, how the sound-words of geometry – produced and heard by means of physical processes – came to be tied up in a network of bodily experiences that the sound-words eventually denoted in a metonymic way, that is, as parts that denote the whole (Roth & Thom, 2009a). In this chapter, we further develop this framework by showing how the living body mediates between and integrates sound-words and other experiences in the process of conceptual development.

More than any other psychological theory of development, Vygotsky's (1934/1986) approach to a child's conceptual understanding is distinguished by the explicit attention it gives to the dynamic relation between thought and language and to the generative development both undergo. In some educational studies, literacy is treated as a problem of words, numbers, and (mathematical) representations that are independent of concrete (bodily) acts that is, from the living sensuous labor that generates, interprets, and transforms those linguistic products. This assumption about (mathematical) literacy constitutes a justification for written tests at school, nationwide, or at international levels. The separation of participative (unindifferent) understanding from the bodily act, sound, or communication – which Vygotsky or Bakhtin theorize as but two aspects of the same unit – deletes and mystifies the dynamic relation of the sound-word to thought and therefore leads to explain conceptual understanding in an inexplicable, mysterious way. Here we use the term *sound-word,* which focuses attention on the fact that words really are physical sounds that need to be produced by the speakers' physical bodies and that are received and translated by the hearers' physical bodies (ears). They initially and primarily do not serve representing function but orienting one another in a familiar world (e.g., Wittgenstein, 1953/1997).

In this chapter, we theorize the dynamic development of participative (unindifferent) understanding while building upon our previous work on sound-words as metonymic moments of mathematical concepts. We attend to the dynamic that to the child the sound-word initially is "an integral part of the object it denotes" (Vygotsky, 1934/1986, p. 222) and subsequently changes to being a functional "tool" at the decisive instant in concept formation. We exemplify two assertions about the role of the body in communication that exhibit the living relation between thinking and speaking in the articulation of the child's "verbal thought" and its development. We exhibit the vital role of the body in the generative development of verbal thought, which comes to be performed and supported by different corporeal means that are produced in concert.

DEVELOPMENT OF CONCEPTS: THE BODY AS A MEDIATING HUB

A child's development of conceptual understanding is dynamically intertwined with the development of participative (unindifferent) understanding in communication (Vygotsky, 1934/1986). Writing on the method of studying children's scientific conceptions, Vygotsky criticizes research on conceptions that tends to "concentrate on the word," which therefore deals with the finished product of concept formation" and "overlook[s] the dynamics and the development of the process itself" (p. 96). On the other hand, this research is concerned with the psychic processes abstracted from perceptual materials and therefore "disregard[s] the role of the word in concept formation" (p. 97). The separation of the word and the sensuously lived signification, which related the child to the world, is built on the assumption that (mathematical) conceptions consist of cognitive processes that are independent of practical activity and the corporeal experience with objects (materials, representations, language).

Our body-centered framework of literacy and participative (unindifferent) understanding extends the Vygotskian approach to the child's conceptual development in two ways. First, the body is central to the *living and lived* articulation of verbal thought, which comes into existence only in and through the expression (i.e., according to Vygotsky and Merleau-Ponty, thought does not exist prior to its expression). Previous studies on corporeal knowing show that the bodily (material) experience *with* objects (world) constitutes the ground and contents of thinking; linguistic studies show that those corporeal actions can be a constitutive part of communication *about* objects (e.g., body position, gestures) (Roth & Lawless, 2002a). That is, the body centrally mediates the orientation toward the outside world and the dynamic articulation of thoughts. Vygotsky (1934/1986) points out that "[i]n the beginning was the deed" and that "[t]he word was not the beginning – action was there first" (p. 255). The body constitutes something like a hub that makes the inside world and outside world irreducible parts of the same experience: Consciousness of the world and the world that consciousness is conscious of are but two manifestations of the same unit (Leont'ev, 1978). The body constitutes a key phenomenon and category for tracking the origin and development of thought-in-practical-communication, a phenomenon that sublates the distinction between inside and outside.

Second, the body is essential to the real-life development of participative (unindifferent) understanding. Vygotsky (1934/1986) suggests that "concepts . . . cannot be assimilated by the child in a ready-made form, but have to undergo a certain development" (p. 146). Time is an important aspect of this development, which renders "direct communication between minds . . . impossible, not only physically but psychologically" (p. 252). The word is "a living union" (p. 5) of sound and participative (unindifferent) understanding, the production and reception of which the living body makes possible in its temporal and spatial characteristics. In speaking, these characteristics emerge from the physical, phonetic aspects of the sequential sound production; in writing, it is the moving hand and pen that exhibit the characteristics of time and space. In real-time communication wherein the participative (unindifferent) understanding lives and develops, the sound-word and other sense-making resources that the body produces are temporally coordinated. The bodies of speakers and listeners produce the temporality – i.e., rate, changing rates, pulses, rhythms, compressions, or accelerations – that they experience in speaking and hearing the sound-word (Roth, 2011a). The living body thereby mediates the temporal development of participative (unindifferent) understanding in which the child engages as it takes part in sympractical activity. In what follows, we take this perspective and articulate a theory of the conceptual development in mathematics. We present and support a theory summarized by two assertions on the role of the body in conceptual development: (a) in communication with an Other, the body translates between the experience of the world (objects) and the sound-word and (b) the body constitutes something like a hub that generates a new unit of participative (unindifferent) understanding.

Between World and Communication

In this section, we show how, in communication, the living body mediates between the experience of the world (objects) and the sound-word. We analyze the dynamic of this simultaneous (i.e., instantaneous) translation between the experience of the world and word in communication with a conversation from the same classroom that also features in chapter 1. In the present episode, the teacher and her group of elementary school children discuss the shapes of three-dimensional objects in a geometry lesson (Figure 4.1).

The children participate in a classification task and thereby learn about geometric categories and representations (Roth & Thom, 2009b). For children who are engaged in sorting out three-dimensional objects but do not yet know their final outcome, communication constitutes a (developmental) process by means of which they encounter various (material, linguistic) resources of mathematical categories and mobilize them as part of the trajectory from their current ways of knowing objects to the future ways. The category is inseparable from the societal setting, here the school, in which children talk in reference to concrete objects and engage in classifying them. Therefore, children's development of mathematical concepts in classification (i.e., abstraction) pertains to the living labor that translates the perceptually salient properties of things to categorical articulation in communication. The body relates the (material, perceptual) experience of mathematical objects and verbal articulation. Immediately, the mobilization of a categorical language in communication constitutes and affects the way by means of which children experience and know mathematical objects.

Transcript 4.1

01 Teacher: [what about this one]
[((*The teacher grabs a pad of sticky notes from away and puts it on the floor*))]
02 [(0.9)]

Figure 4.1. A group of geometric objects is placed on an A4-sized piece of purple-colored paper for a whole class discussion.

[((*Evan puts down another pad that he holds on his right hand and grabs the pad. Evan's left hand is put on the top of a wooden block placed on the floor*))] (Figure 4.2a)

03 Evan: [<<p>well>, this one's]
[((*Evan picks up with his right hand the pad that the teacher brought. Bill's eye gaze follows Evan's hand movement*))] (Figure 4.2b)

04 [not really like a square]
[((*Evan slightly takes away his left hand from the topside of the wooden block and grabs it again. *He picks up the block and holds it. The teacher stretches the yellow cube toward the students*))] (Figure 4.2c)

05 [because it doesn't match the square like this]
[((*Evan puts the two objects together. Bill and Clara gaze at them. Evan turns the attached side of the two objects toward the teacher. His gaze changes from the blocks to the teacher. *Bill tilts his head and body*))] (Figure 4.2d)

06 Teacher: okay but it's- [why is it why did you]

07 Clara: [I didn't (find out?)]

08 Teacher: [why did you pick them up]
[((*The teacher moves her left hand near the two objects that Evan holds. Bill moves up several pieces of sticky paper on his hands*))] (Figure 4.2e)

09 [why are they kind of like a cu:be]
[((*Bill gazes at the pieces of papers on his hands. The teacher stretches her left arm to Bill and takes the paper that he holds. *She brings the yellow cube toward students*))] (Figure 4.2f)

Description

The whole class sits in a circle on the classroom floor and talks about the geometrical objects that Clara, Evan, and Bill have collected by considering those to belong to the same kind with the sample (i.e., one of the mid-size wooden cubes in Figure 4.1). The teacher walks toward the three students and crouches down next

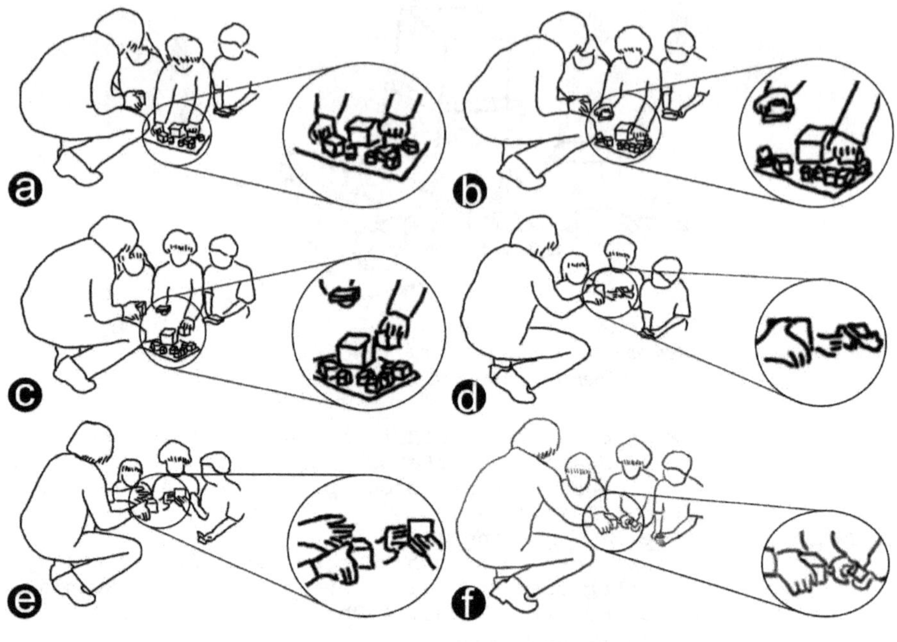

Figure 4.2. a. Evan grabs with his right hand the pad that the teacher put on the paper. His left hand is put on the top of a wooden cube. b. He picks up the pad with his right hand. c. Evan picks up the wooden cube with his left hand. d. He puts together the pad of sticky notes and the small wooden cube and gazes at the teacher. Bill tilts his head and body and gazes at the two objects from the side. e. The teacher moves her left hand near the two objects that Evan holds. f. She stretches the cube on her right hand toward the students.

to them (Figure 4.2). All the while, she holds a yellow cube in her right hand. "What about this one," she says and grabs with her left hand a pad of sticky notes that is located on the purple-colored paper beneath (line 01). She moves the pad and puts it on the floor near the three students. Evan returns another pad and instead grabs the pad that the teacher has moved to his front (line 02, Figure 4.2a). He suggests that the pad is not really like a square (lines 03–04). Evan picks up the pad of sticky notes (line 03, Figure 4.2b). Bill gazes at it following Evan's hand movement (line 03). Evan grabs a cube-shape wooden block with his left hand, which is one of the mid-size blocks on the purple-colored paper (Figure 4.1). He slightly moves his hand up and away from the wooden block and eventually picks it up (Figure 4.2c). Simultaneously the teacher stretches her right arm and brings the yellow cube on her right hand toward the students (line 04).

Evan holds the pad of sticky notes on the right hand and the wooden block on the left hand. He puts them together (Figure 4.2d). He says that the face of the pad does not match the face of the wooden cube that is a square (line 05). Clara and Bill gaze at the two blocks that Evan holds. Bill slightly tilts his head and gazes at the two blocks from the side (line 05, Figure 4.2d). The teacher says "okay but it's" and goes on, "why is it why did you" (line 06). Clara's utterance overlaps the teacher's (line 07). The teacher continues, "why did you pick them up" (line 08). She brings her left hand on its palm facing the blocks that Evan holds, and thereby makes light waving gestures (line 08, Figure 4.2e). The teacher asks why these are kind of like a cube (line 09). She stretches her left arm to Bill and takes two individual pads of sticky notes that Bill holds. She takes them away and reaches the yellow cube in her right hand to the students (line 09, Figure 4.2f).

Analysis

The episode exemplifies the central role of the living body with respect to which the world (objects) takes shape and which, simultaneously, is structured by the world (objects). In this situation, a teacher and her students talk about the geometrical properties of the objects that Clara, Evan, and Bill have collected and compares it to a categorical sample. The teacher grabs a pad of sticky notes placed away from the purple-colored paper and brings it forward to the students. "What about this one," she says and puts it on the paper (line 01). Evan grabs the pad that the teacher has brought by putting down an object (another pad of paper) that he has already held on his right hand (line 02). He immediately holds up the pad and utters "well, this one's" (line 03), literally repeating the words the teacher has said before ("this one" [line 01]). Evan thereby takes his communicative turn with his hand movement and speech. His bodily action produces an effect that establishes the teacher's utterance ("what about this one" [line 01]) and hand movement (moving a pad of sticky notes [line 01]) as a salient question that seeks to be addressed.

Evan pushes his communicative turn and answers the teacher's questioning. He utters "not really like a square" and simultaneously holds up a wooden block, on the top of which he had put his left hand previously (lines 03–04). The wooden block is an all-square-faced sample (i.e., cube) that the three children were given as a reference object. Although real, it takes the place of and denotes the ideal objects that geometry is about (Husserl, 1939). The children have gathered objects to build a collection that came to constitute their category. Therefore, Evan's verbal articulation of the word "square" and his corporeal action of holding up the wooden block followed by the pad respectively with his left and right hand, make salient a particular geometrical property of the pad: the shape of the face of the pad in comparison to the shape of the square. Evan's body mobilizes a concrete reference object that contributes to marking the sense of the word, an experience that he verbally and metonymically indexes by saying "square." Those bodily actions thereby relate the object of communication from "this one" (line 01) to "not really like a square" (line 03), which constitutes a form of abstraction denoted by the sound transcribed as [skweə] according to the International Phonetics Alphabet, a sound-word that English speakers hear/understand as "square."

The example shows that the translation (abstraction) involves a concrete instant in which the body dynamically integrates some of the structural resources available to both speaker and audience. The corporeal mobilization of the wooden cube joined with the previously established object of communication (i.e., the pad of sticky notes) constitutes the actor's involvement in the ongoing categorical interpretation. This concretely integrates the presence of the wooden cube in the collection and the presupposed understanding (i.e., sample) about its belonging there. Just as Sartre (1943) suggests: the body and the world mutually ensnare each other. This body does not exist of disparate elements that some mind needs to relate but it is the seat of an "I can" that holds all of these experiences together in one unit (Henry, 2003). The action of holding up the wooden block constitutes an irreducible component of the interpretive movement toward "square." This living body constitutes the means that concretely realizes the condition that makes this communication possible (i.e., the presence of blocks as the object of classification) and that generates the category of the object. It is in the body as flesh that a specific aspect of the geometrical properties of the pad comes to be sensibly marked with the sound-word – the sense of "this one" is marked with the sound-word ("not like a square") and the lived perception of the sample cube.

The living body, the flesh, is the one locus where the sound-word and the other associated resources are pulled together and linked. This affects the body and the way it experiences the objects. Thus, Evan articulates "it" as not matching "the square like this" (line 05). He puts together the pad and the wooden cube and turns the wide side (face) of the paper pad toward the teacher. Evan begins with a "because" and thereby grammatically indicates that the later part of his utterance will provide some reason for his preceding utterance. This articulation of the reason ("doesn't match the square" [line 05]) and his action of physically joining the two faces of the two objects explicate the particularities (geometrical properties) of the pad. They generate the specific form of geometrical configuration and thereby realize a more developed form of the physical encounter with the pad and the wooden block ("doesn't match"). The body is ensnared by a more developed categorical experience ("match" [line 05]), which opens up in the course of the previous articulation of the word "square" and the mobilization of the two objects. The body becomes the locus of a relation in the course of speaking. The communicative resources mobilized in this course lead the speaker to a new form of experience.

Since sound-word and the configuration of geometric objects are mobilized in the public space of communication, the translation of the experience brought about by the mobilization of objects pertains not only to the speaker but also to the audience. It is not only co-present in the conversation and listening to the talk but in fact constitutive of each utterance, each word (Bakhtine [Volochinov], 1977). We observe the development of the corporeal experience of the objects in the subsequent action of Bill, who sits next to Evan. Bill turns his head and body and gazes at the newly joined objects from the side. His movement exhibits his changing attention toward the matching configuration of the two faces and therefore the possibility of the different perception made possible by his changed body position, orientation, and the angle of his eye gaze. His body is located in a discursive space in which the body is open to be affected by the ongoing talk and where it encounters new, not-yet-realized possibilities. In this space, the living labor integrates the categorical experience of the geometric objects and the articulation of the (categorical) word. These affect one another and therefore lead to the development of the whole unit that we denote by the term *participative (unindifferent) understanding.*

This example exhibits the lived, *collective* labor of teacher and students. Because this labor is collective, the teacher participates in the students' production of connections between sound-words and other bodily experiences. According to Vygotsky (1989), it is precisely this very "physical relation between people" (p. 56) that *is* the higher psychological function that one can subsequently find operating independently in the child. It is not that this function first exists between the co-participants and then within the child, because in participating, the child already contributes to producing the higher psychological function. In the turns 06–09, the teacher responds to Evan and pushes them to ongoing verbalization rather than simply asking them to complete the classification. The teacher begins by uttering "okay but it's why is it why did you" (line 06). This utterance comprises grammatically incomplete phrases and repetitions ("it's," "why is it," and "why did you"). She then further translates her initial utterance (line 06) to "why did you pick them up" (line 08). She moves her left palm near Evan's two objects and waves at them. The hand movement thereby underlines the speaker's attention to those objects. All the while, she holds a yellow cube in her right hand and near those objects. Bill changes his gaze to the two individual pads of sticky notes that he has been holding in his hands throughout the conversation. Bill slightly moves them up and down and gazes at them. The teacher stretches her left hand to grab those pieces of papers and takes them away. Instead she holds her sample yellow cube in the right hand and tends it to Evan and Bill. "Why are they kind of like a cube," she says (line 09). By stretching the "u," she elongates the sound-word that we hear as [kju:::b]. The teacher's questioning thereby makes the categorical sound-word ("the cube") stand out and against the material example (the yellow cube) together with the objects previously mobilized (the pad of sticky notes, the wooden cube). This physical, phonetic stretching of the sound and the cognitive emphasis alone would be sufficient evidence for the coincidence of body and mind, their "undecidable" difference.

Discussion

This analysis exemplifies the generative role of living labor in the sequential coordination of objects (Figure 4.3). The initially given seed of an idea for a categorization grows in to a more developed unit in and through a series of developmental translations. The sequence therefore pertains to the child development of categorical understanding in communication. The analysis shows how the involvement of the body in the world concretely realizes the metonymic relation between the sound-word and the totality of sense-marking and sense making resources available in situ, including body movement, gestures, eye gaze, body position and orientation. The living body opens up to be affected and constitutes a place where the co-appearance of these different experiences can be marked (and therefore remembered). On the one hand, the body constitutes the source of an individual's position and orientation in communication, and thereby mediates an individual's body-

["this one"|((Teacher's left hand moving a pad of sticky notes on the floor))]
↓
["this one"|((Evan's right hand holding up the pad))]
↓
["not really like a square"|((Evan's left hand holding a wooden block))]
↓
["doesn't match the square like this"|((Evan's hands joining the pad and the block))]
↓
["them"|((The teacher moving her left hand near the objects that Evan holds))]
↓
["they kind of like a cube"|((Teacher's right hand bringing forward the yellow cube))]

Figure 4.3. The sequential translation of the communication unit is represented in a bracket by the combination of sound-words and other bodily sense making resources. The Sheffer stroke "|" denotes their irreducible and mutually constitutive relation.

centered, positioned, and unique (singular) way of knowing the world. On the other hand, the body constitutes this singularity as it occupies a specific location in space-time. Any individual movement and recorded corporeal experience can serve as the metonym for the total experience not only for the Self but also, and irreducibly so, for the Other. The living labor of translation makes the child experience the sound-word differently than before and, in the process, develops participative (unindifferent) understanding. This, we understand, exists in the metonymic relation between the sound-word and the totality of relevant contexts that it indexes. The teacher in communication contributes to the child's conceptual development by increasing the possibilities for translation.

REALIZING NEW PARTICIPATIVE (UNINDIFFERENT) UNDERSTANDING

In this section we show how the body constitutes something like a hub that realizes a new participative (unindifferent) understanding. The relation of the sound-word to the thing-in-context that it marks in communication is a dynamic one. In the live performance of a classification of objects, the categorical sound-word (e.g., square) and other resources that mark the sense of the category mutually affect one another and translate the geometrical constitution of objects. The body mediates the development of participative (unindifferent) understanding by concretely realizing the sound-word as part of a new categorical experience of the world (objects). As the body mediates the constitution of a communicative unit that marks the sense of the category differently than before (because the sound only indexes an ever developing network of experiences), the child's verbal articulation and the associated conception (participative [unindifferent] understanding) develops. In this development, objects enter into a different relation with the body. For example, objects enter into the unmediated, direct relation with the body and gain a functional role in the articulation of a category. In this section, we analyze a situation that immediately followed the episode featured in the previous section. It exemplifies the fundamental role of the body in the constitution of a more developed unit of participative (unindifferent) understanding.

Transcript 4.2

16 Teacher: [Bill]
[((*The teacher points at Bill with her right hand*))]
17 [(0.6)]
[((**Bill stretches his right arm toward a pad of sticky notes outside the purple-colored paper. The teacher bends and stretches her right hand and brings a yellow cube to the collection of objects*))] (Figure 4.4a)
18 Bill: [because um]
[((*Bill picks up the pad. The teacher puts the yellow cube on the top of the red block*))]
19 (0.7)
20 this um ((**Bill tilts his head and body and gazes at the pad on his hands from the side*)) (Figure 4.4b)
21 [this um]
[((**Bill further tilts his head and body*))] (Figure 4.4c)
22 [probably if they had]
[((*Bill grabs the yellow cube with his right hand. He moves it aside a little to the right and thereby makes a space over the red block. *He brings the pad in his left hand to the space from the left side*))]
(Figure 4.4d)
23 [more um]
[((*Bill puts the pad down on the top of the red block*))]
24 little paper it [(wide?)]
[((**Bill lightly touches the upper side of the pad*))] (Figure 4.4e)
25 [actually]
[((Bill moves up the hand as high as to the top side of the yellow cube))]
26 [fill up to there]
[((**Bill holds the hand there and gazes at the teacher*))] (Figure 4.4f)
27 [and if (??) a little bit wider]
[((**Bill brings his hands to the front and the back side of the pad each. He pulls the pad and lines it up with the foreside of the yellow cube*))] (Figure 4.4g)
28 ((**Bill keeps the position of the pad and moves away his right hand almost to the back side of the yellow cube*)) (Figure 4.4h)

Description

The teacher utters Bill's name and points at him with the index finger of her right hand with which she holds the yellow cube (line 16). Bill immediately reaches out and grabs the pad of sticky notes (line 17, Figure 4.4a). The teacher bends over and stretches her right hand toward the collection of objects on the purple-colored pa per (line 17). Bill takes the pad to his front and utters "because um" (line 18). Simultaneously the teacher puts the yellow cube on the top of the red block. Bill

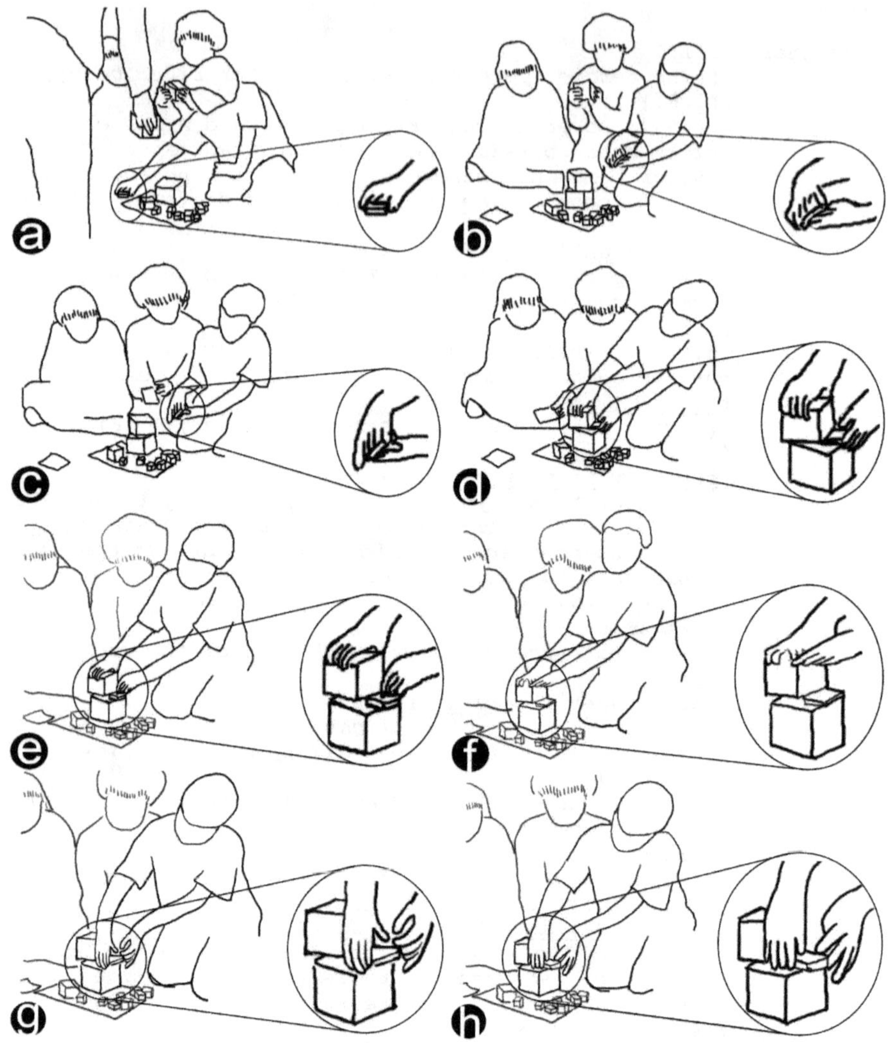

Figure 4.4. a. Bill stretches his right arm and grabs the pad. The teacher stretches her right hand and brings the yellow cube toward the other blocks on the floor. b. Bill holds the pad with his hands and tilts his head and body to the left. c. Bill further tilts his head and body toward the left side of the blocks. d. Bill moves the yellow block to the right with his right hand and puts the pad on the top of the red block. e. Bill grabs the yellow cube with his right hand and lightly taps the topside of the pad with his left hand. f. Bill moves up the hand as high as to the top side of the yellow cube and gazes at the teacher. g. Bill grabs with his hands the front and back side of the pad. He pulls the pad to his front and lines it up with the foreside of the yellow cube. h. He holds the pad and moves his right hand farther away from the back side of the pad.

gazes at the pad and utters "this um" (line 20). He slightly tilts his head and body and gazes at the pad from the side (line 20, Figure 4.4b). Bill then moves the pad close to the blocks on the paper and repeats "this um" (line 21). He further tilts his head and body (line 21, Figure 4.4c). Bill utters "probably if they had" (line 22). He moves the yellow block a little to the right, and thereby makes a little space on the top of the red block. He brings the pad of sticky notes into his personal space (line 22). Bill puts the pad on the topside of the red block right beside the yellow cube (lines 22–23, Figure 4.4d). He says "if they had more little paper" (lines 22–24). Bill lightly taps the upper side of the pad with his right hand (line 24, Figure 4.4e). He moves the hand up as high as to the topside of the yellow cube (line 25). He utters "actually" (line 25). Bill holds his hand at that height, which we may understand as if he had said "fill up to there" (line 26). He orients his gaze away from the blocks and to the teacher (line 26, Figure 4.4f). Bill holds his left hand at the front side of the pad and his right hand at its backside (line 27, Figure 4.4g). Bill says "a little bit wider" (line 27), which we may hear as indicating the condition of expanding the width of the sticky notes. He moves his right hand backwards (away from him) until it reaches almost the backside of the yellow cube from the back end of sticky notes (line 28, Figure 4.4h).

Analysis

The episode exemplifies the generation of sense-making resources that constitute the development of categorical understanding. Bill mobilizes the pad of sticky notes and speaks beginning with "because," which thereby produces the answer to the question that the teacher had asked in the episode of the previous sub-section (line 09). First, Bill moves up his left hand almost as high as the top of the yellow cube in his right hand (Figures 4.4e, f). Second, he puts his hands as wide as the sticky notes and then moves the right hand a bit farther almost to the end side of the yellow cube (Figures 4.4g, h). Bill's movements and associated utterances posit stacking more pieces of paper and expanding their size. Those actions thereby extend the shape of the pad and constitute a three-dimensional iconic figure that is as high and wide as the yellow cube right beside. They transform the physical shape of the pad into a three-dimensional object that has square-shaped faces. They constitute a presentation of a cube that can be seen to denote the existing one; he thereby produces a translation of the pad to a cube. Therefore, Bill's utterances and other body movements concretely articulate the geometrical properties of a cube in the conversational space. A new signification for the categorical word "cube" arises from the resources that are produced and coordinated around the pad and the yellow cube.

The episode exemplifies the constitution of a new form of participative (unindifferent) understanding and therefore also the development thereof: how the development occurs in and through the living labor in a world populated with objects. The new participative (unindifferent) understanding arises from the real-time generation and constitution of sense-marking resources that the living body produces in this place and at this time. For example, Bill mobilizes the yellow cube as part of his talk, which the teacher returns to the collection of blocks only after Bill picks up the pad and begins his talk. The yellow cube does not constitute a perceptually salient object at the instant when Bill is called and when he grabs the pad of sticky notes. Therefore, the yellow cube becomes integrated into Bill's talk in real-time as he sees it being put on the red cube and puts it aside to make a space for the pad. Bill puts the sticky note pad and the yellow cube side by side not because he has planned to do this but because he is in the process of coping with the unfolding material condition in which he currently finds himself. The presence of the yellow cube besides the sticky note pad constitutes the structural resource that marks sense, from which emerges the proposal of stacking more paper and thereby increasing the height of the pad. This transformation of the height constitutes the new condition that Bill expands by proposing to increase the width of the pad. The new connections arise from the dynamic relation of Bill's corporeal engagement – with Sartre, we might say ensnarement – in the world (of objects) and the mundane knowing-in-process that is articulated in and through this engagement.

Our dynamic approach to a body-centered nature of participative (unindifferent) understanding shows the significance of integrating the capacity of the living body in the theoretical framework of development. The living body is open to the world – open to be affected *by* the world – and therefore endowed with the power to augment the relations between various experiences, including the speaking and hearing of sound-words. It is thereby enabled to generate new forms of participative (unindifferent) understanding. Thus, the "I can" that we mark and denote by the name "Bill" mobilizes the pad and the yellow cube as an integral part of his living labor by using them as a taken-for-granted (unattended-to) part of the world. In Bill's actions, the salient object at hand no longer is the individual pad or the yellow block but the new conceptual totality that he realizes corporeally. The relation of the body to the pad, for example, changes from the objectifying and objectified relation exemplified in the first episode to the living relation observable in the second. The pad is found in a different relation to the living body and thereby becomes constitutive part of a newly evolved participative (unindifferent) understanding. Therefore, the emergence of a new form of understanding is dynamically intertwined with the change of perception. The speaker and the audience are given the interpretive resources that allow perceiving the materials (the pad, the yellow cube) from a particular (geometrical) perspective that differs from the perspectives they could take before.

Discussion

This analysis provides an example of the development of participative (unindifferent) understanding that appears in the course of classifying geometrical objects and particularly in talking about an object not-yet-categorized (the pad of sticky notes). In learning geometry, a categorical sound-word (e.g., [kju:b] "cube") is a resource for better understanding the totality of produced and transformed significations. A sound-word thereby *accrues to, becomes a constitutive part of, a totality of significations* available in the situation rather than gaining something that comes to be attached to it (like the "meaning" other scholars suggest students "construct" for a word). This sound-word is useful in this situation and the question whether it is useful and used elsewhere cannot be presupposed but needs to be taken as an empirical matter. The conceptual power of a geometrical category depends on the whole unit in which the sound-word is only *one* among many constitutive parts. The example shows that the participative (unindifferent) understanding of a category-in-situation involves more than the use of the sound-word (e.g., [kju:b] "cube") for making a simple correspondence between it and an object (e.g., a physical CUBE). The resources that Bill produces in communication may be used subsequently for justifying either the sameness or the difference between the pad of sticky notes and the sample cube. The example shows how this corporeal relation of the particular objects at hand produces the conceptual gains that constitute the development. The learning of mathematical classification exists in the increasing capacity of the learner to activate a category in a variety of ways – sound-word, iconic gestures, pointing gestures, body orientations, and so on.

IRREDUCIBILITY OF PARTICIPATIVE (UNINDIFFERENT) UNDERSTANDING

In this chapter, we extend previous work on the nature of mathematical conceptions by showing how new categories arise from and are tied to various bodily presentations. Linking our work to a close reading of Vygotsky's *Thought and Language* concerning the development of concepts, we exemplify two facets of the emergence of participative (unindifferent) understanding: (a) the body as the locus and organ where different experiences of the world (objects, sound-words) come to be interconnected and (b) the body as the locus and organ that realizes a new form of participative (unindifferent) understanding. Our work here implies that categorization reduced to making a mechanical correspondence between the word and the thing is like studying hydrogen and oxygen to know the properties of water. In the analysis of conceptual development, deleting the "body" is like deleting "time" from the logic of practical, everyday experience.

This chapter exemplifies how in real (classroom) instruction, the development of students' conceptual understanding is generated by the real-time linguistic interaction that does not necessarily have to reproduce an instructional model – either predesigned or constructed afterwards – in the exactly same way. As teacher and students communicate while participating in classroom events, the words they use are not their own – any speaker borrows the sound-word that always and already has come to her from the Other and, in speaking, returns to the Other. Even in those instants when someone creates a new sound-word, it is, as word, always and already accepted as something intelligible. It would not make sense to offer a sound that has no sense or function. Words yoke speakers and listeners together into the same unit (Bakhtine [Volochinov, 1977), which we refer to here as participative (unindifferent) understanding. Any sound-word marks significations that vary in the course of developing participative (unindifferent) understanding. The very condition that makes communication possible in and for teaching/learning (i.e., the plurality of the word) simultaneously involves the possibility for bifurcation and therefore different trajectories of actually realized participative (unindifferent) understanding. Language and literacy thereby constitute not merely a medium that delivers a single signification of something else (e.g., an instruction model) but that is itself a dynamic, intelligible process that generates the development of participative (unindifferent) understanding.

Our work elaborates on conceptual seeds initially sown by Vygotsky. To him, literacy constitutes an integral part of the child's development of scientific (mathematical) concepts in that (a) the sound-word mediates conceptual understanding (i.e., thinking) and (b) the participative (unindifferent) understanding undergoes continuous development. In this framework, participative (unindifferent) understanding is, as exemplified here, a dynamic process rather than a thing; this process is genetically inseparable from the corporeal act (e.g., gestures) that appears along with the sound, which itself is the result of bodily movements (vocal cords). An advanced way of theorizing the child's development of participative (unindifferent) understanding (i.e., conceptual understanding) is therefore found in the role of the body. Thus, in communication, my living body constitutes the singular organ that brings about a particular trajectory of my participative (unindifferent) understanding. My body allows me to speak and hear a sound-word in a particular way (e.g., prosody, body orientation). Simultaneously, the fact that the sound-word is not just my own (i.e., singular plurality) raises the need to theorize the body as an open space, that is, as a hub that irreducibly articulates the inner and outer of the social-psychological being. In our approach, it is the openness of the living body to the world that allows it to be shaped in and through the social relations that are, as Vygotsky suggests, the origin of *any* higher cognitive function. It is precisely because of my living body that I can reproduce and transform cultural-historical participative (unindifferent) understandings and therefore contribute to the reproduction of geometry (algebra, physics, chemistry, etc.) as objective science.

CHAPTER 5

MATHEMATICAL INSCRIPTIONS AND CULTURAL DEVELOPMENT

> It is the important function of the written to enable the continual objectivity of ideal sense formations in the curious form of *virtuality*. Once documented in writing, the ideal object exists virtually in the world and may be produced actually at any instant. (Husserl, 1939, p. 212)

Written mathematical forms – i.e., mathematical inscriptions – not only constitute the very essence of science but they also produce its objectivity. Scientific inscriptions (e.g., force diagrams, chemical reaction equations, DNA models), too, constitute resources for mediating the understanding of science concepts. This is so because, as Vygotsky (1989) suggests, "man controls the activity of his brain from without through stimuli" (p. 59). It is not surprising, therefore, to find that teachers mobilize in their lessons different types of mathematical inscriptions to explain science concepts. Inscriptions are inherently material. In communication, these are connected to other sense making resources (e.g., language), and thereby come to stand for something (i.e., mathematical concept). Yet, students have difficulty reading mathematical inscriptions that they encounter in science lessons and associated talk. On the other hand, the very historicity of a science, its continual development in time as objective science, is enabled by the inscriptions: "How the tradition of sense formation of elementary concepts proceeds may be seen in elementary geometry curriculum with its methods and textbooks. What we really learn there is the use of finished concepts and statements in highly methodical ways" (Husserl, 1939, p. 217). That is, learning science and mathematics involves a contradiction. The objects that students deal with are complex because they embody the accumulated results of mathematical and scientific experiences. And yet, the possibility to accumulate experiences in complex artifacts enables cultural development without requiring that all preceding experiences have to be relived by the members of new generations.

The difficulty of learning mathematics and science that arises from this contradiction tends to be attributed to students' individual (mental) capacities or the qualities of mathematical representations, which are said to exist independently from the concrete praxis of communication. In this chapter, we continue to develop a holistic approach to understanding, here the understanding of mathematical inscriptions. Our approach does not dichotomize sense making and the sensuous experience of the world (objects) and therefore leads to implications for the pedagogical problem of "representation" that differ from other epistemological theories used in science education (e.g., constructivism). We thematize the dynamic experience of mathematical inscriptions in communication, which we summarize into three assertions. First, the body reproduces and transforms cultural resources for translating mathematical representations. Second, the increase of heterogeneous sense making resources in communication increases possibilities for realizing new ways of talking. Third, knowing mathematical inscriptions emerges from an irreducible unit that integrates all communicative modalities into one heterogeneous ensemble. We illustrate the three assertions by analyzing a case in which children talk about an arbitrary geometrical configuration in a mathematics lesson. We conclude that knowing mathematical representations is equivalent to corporeal reading between different sense making resources – in the sense of "reading between the lines" – that constitute a series of references to bring about scientific conceptions in their totality.

PROBLEM OF REPRESENTATION

Mathematical forms constitute pervasive sense making resources in the pedagogical practice of science education (e.g., diagrams, geometrical models, or graphs). Science teachers draw on mathematical inscriptions as part of their communicative interactions with students. Despite the significance of mathematical inscriptions and associated studies on learning, pedagogical theories related to the use of mathematical inscription have been less informed by learning theories. Rather, prescriptive teaching models describe learning from a perspective that is constructed after inscriptions are already known. Mathematical inscriptions tend to be treated as if they possessed significations ("meanings") stably attached to them and should express participative (unindifferent) understanding self-evidently. This assumption is exemplified when a (beginning) teacher attempts to present a number of mathematical inscriptions in a short talk and does not provide sufficient amount of time to students for reading them. It is well known among experienced teachers that the simple introduction of nice inscriptions does not necessarily promise students' successful understanding of scientific ideas. One teacher's effective use of a mathematical inscription does not promise another teacher's success with it in another situation. This is so because mathematical inscriptions constitute only one part of a communicative whole that in its entirety stands for a scientific or mathematical idea. The reason for a successful or unsuccessful lesson may not be fully explained by the use or non-use of a specific inscription but requires the consideration of sense making resources simultaneously mobilized in communication. Therefore, the purpose of this chapter is to exhibit the ensemble of different sense making resources that dynamically interrelate to signify some idea. We aim at understanding lived experiences of mathematical forms in real time and articulate pedagogical principles theorized from a perspective of students who do not yet know mathematical inscriptions and therefore cannot intend to learn.

From a pedagogical perspective, knowing mathematical inscriptions involves a dialectic problem. Students encounter new mathematical inscriptions (e.g., force diagrams) and depend on teacher's talk and other textual references to integrate them into the familiar lived-in world. It is only through its integration into a network of signification that an inscription becomes intelligible – a people without contact with Western culture would not use a mobile phone that had fallen from the sky in the way we know it. However, teachers' talk with/about mathematical inscriptions, for example, makes sense to students only when they are already familiar with aspects of the world that those inscriptions refer to. This is the case, for example, when I read a "difficult" book and ultimately say that I find nothing familiar or when I listens to some science or mathematics talk and hear nothing but "gobbledygook," that is, I find it sounding like something said in a foreign language.

From a linguistic perspective, understanding mathematical inscriptions involves relating one material body (i.e., a graph) to another material body (e.g., a tree). This is possible because of a series of connections that an inscription makes with other aspects of culture. Mathematical inscriptions mark some sense to readers when there is a web of cultural significations available. Yet, the mutually presupposing relation between mathematical inscriptions and sense making resources indicates that cultural significations consist of an infinite number of relations. These are stabilized at some stage of communication (Eco, 1976). For example, a science teacher explains Newton's second law by presenting a mathematical equation – $\underline{F} = \mathrm{m} \cdot \underline{a}$ – followed by analyzing a constant-velocity motion graph in a zero-net-force situation. He might explain the concepts of inertia and acceleration, enact some demonstration, and so on. Still, the teacher has to make a decision whether to provide additional sense making resources. The potentially infinite number of sense making resources required for setting up something such as a inscription raises a question about the very possibility of reading (understanding) mathematical inscriptions: In what ways do people produce the large number of relations required in the understanding of a culturally existing idea? How do students come to be able to read from a force diagram the non/equilibrium of multiples forces acting on an object and the resulting motion of the object? Since knowing mathematical inscriptions (e.g., force diagram) means being able to enlist appropriate sense making resources (e.g., configuration of real objects, verbal description) and therefore let one material body stand for some other (i.e., making a signifier-signified relation), the amount and extent of sense making resources that students need turns out to be an empirical matter. It depends on the concrete ways by which people experience the world and find their ways (i.e., culture). The role of pedagogical practice is to assist students in coming to know (read) inscriptions and understand scientific (mathematical) ideas involved in them.

We approach the pedagogical problem of mathematical inscription by taking a holistic approach to communication, which allows dealing with the problem by addressing the cultural, linguistic, and corporeal nature of learning rather than making it a matter of individual knowing ("construction"). That is, we take a perspective that does not theorize thinking as a function of language (which would literally be *logo*centrism) but consider the different communicative modalities as one integrated whole. This whole manifests itself differently and one-sidedly in the modalities, none of which therefore can represent the whole though each certainly suffices to point us to the whole. In what follows, we articulate a Vygotsky-inspired theoretical framework of knowing mathematical inscriptions. We draw on the phenomenological concept of the living body as the source of sense. We exemplify the significance of this approach by analyzing a process by means of which an arbitrary geometrical shape, here inscribed in material form, first comes to be known to children in and through communication.

THE LIVING BODY AS CULTURAL SIGNIFICATION

> That our knowledge, the present and living cultural form "geometry" is tradition and simultaneously a form of handing down, is not merely knowledge about an external causality that brought about the sequence of historical figures but an understanding of its true nature of being. (Husserl, 1939, p. 220)

In the living praxis of teaching and learning, mathematical inscriptions constitute communicative resources mobilized together with other forms of corporeally produced communication – such as words (verbally spoken), gestures, body positions, and orientations. As a speaker, I display or hand-write a specific form of inscription (e.g., force diagram) and might simultaneously talk, point out some part of it, and change a position or orientation that I take up with respect to it. Mathematical inscriptions constitute a sense-making corpus (i.e., bodies) toward which teachers and students bring their attention and carry out physical interactions. This is so because inscriptions occupy part of the space and could be co-present there for the interaction with human agents. Even in the case of not yet grasping the sense of mathematical inscriptions (e.g., un/balanced multiple forces), students can still point out a part, juxtapose their talk with them (e.g., asking a question, propose a claim), or take up a position beside them. The living body comes into contact with other material bodies (e.g., inscriptions) in the interaction and is located always with respect to those other bodies. In this, the living body comes to be integrated into a network of significations (i.e., signifying–signified relation) and can be marked by the cultural sense that goes beyond individuals' intentions. The notion of the *living* body – or the *flesh* in phenomenological philosophy – indicates this radical capacity of the human body (Henry, 1975). The living body in the active participation in communication always already is caught up in the world where it embodies cultural possibilities and therefore is constitutive part of cultural significations from the beginning. Therefore, the living body constitutes the very possibility for the production of inscription-related practice (i.e., communication) and being part of scientific thinking (i.e., cultural practice).

The living body is not only an individual gesture of thinking but also a(n) (cultural) expression itself, which in communication makes sense not only to the other but also to me. The following four aspects summarize the dynamic role of the living body that mediates cultural significations in the inscription-related communication. First, the body constitutes the locus where the mathematical bodies (inscriptions) are experienced. Second, the body constitutes the medium that produces and is affected (changed) by communication. Third, communication in the presence of mathematical bodies is distributed over a communicative unit that integrates corpo-

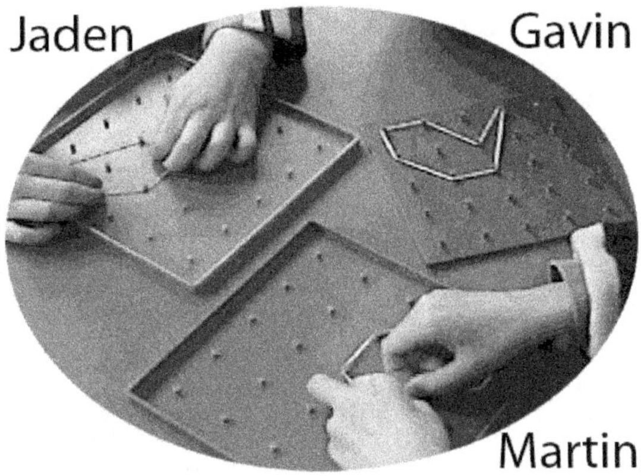

Figure 5.1. Three second-grade boys (Jaden, Gavin, and Martin) sit at a desk in a mathematics classroom. On the table, there is a sample shape (a polygon with a reflex angle) that Gavin produced with a rubber band on a board. Jaden works to make a shape on his board. Martin and Gavin talk to help Jaden.

really produced means (e.g., verbally spoken words, gestures, body movements with respect to inscriptions) with the situation as a whole. Fourth, gestures, body orientations/positions, body movements, which are involved in experiencing mathematical inscriptions, also are involved in communicating them. These four aspects highlight that the entire living body is a signifier of the sense marked by mathematical inscriptions and therefore signified by the culture that it realizes. The living body is an integral part of cultural signification. It unites and reunites sense experiences and sense making in concrete situation of communication in the presence of mathematical inscriptions.

In what follows, we illustrate this holistic approach to mathematical sense making and cultural development around mathematical inscriptions. We thereby exemplify the role of the living body in inscription-related praxis and therefore how it mediates children's cultural development. They are part of the series of generations that reproduce an objective science through a subjectively enabling, performative subjectivity (Husserl, 1939). The empirical example derives from the same second-grade elementary mathematics class that is featured in chapters 1 and 4. By engaging in a variety of experiences, the children are to learn about three-dimensional geometry at a point in time that other theories suggest comes too early in their development (e.g., Roth, 2011a).

The episode is from a group in which three boys are engaged in a conversation about an arbitrary shape that one of them (Gavin) has produced using a rubber band and a plastic board consisting of a matrix of vertical pins (Figure 5.1). The other two boys (Martin, Jaden) work to reproduce the geometry of the same shape on their individual boards. In their conversation reproduced below, Gavin and Martin assist Jaden, who is still working on his configuration. They thereby realize their teacher's suggestion to help Jaden in ways other than actually doing the work for him. That is, the two boys participate in producing a pedagogical discourse the role of which is to bring their peer along to where they are already. The effectiveness of this discourse would be evidenced by Jaden's trajectory of reproducing the geometrical shape successfully while he attends to the communication. That is, the questions of what will be appropriate sense making resources and how they work depends on their role in helping to see and reproduce the *geometrical* structure of the original shape (e.g., the number of sides and corners in specific angles).

We analyze their conversation and exemplify some aspects of the process in which children concretely develop a cultural signification in and through communication. We attend to the role of the living body that is characterized by the capacity to be marked by sense in active engagement with mathematical forms (bodies). We summarize the cultural dynamics of knowing mathematical forms into three assertions to be developed concurrently with our case analysis. First, the body reproduces and transforms cultural resources and connects them to mathematical inscriptions. Second, the sense making resources in communication increase the possibilities for realizing a new way of talking (i.e., a network of signification). Third, knowing mathematical inscriptions emerges from the different, irreducible modes of communication as the integrated whole.

Episode 5.1

```
01  Martin:remember? (look at?)
02         [INSide] [down the-] (?) (look at?)
           [((Martin stretches his rubber hooked onto a pin and
           *moves it back and forth behind the pin))] (Figure
           5.2a)
                    [((Jaden holds his rubber band with his left
                    hand and changes its position from one side
                    to the other of the pin. *Gavin turns his
                    face toward Jaden.))] (Figure 5.2b)
03  Gavin: IN [side]
              [((Jaden holds his rubber band with his left hand
              and *grabs its right part with his right hand.))]
              (Figure 5.2c)
04  Martin:   [Inside (inside?)↑]
              [((Martin moves his rubber band back and forth five
              times behind the pin. *Jaden stretches the right
              part of his rubber band around the pin.))] (Figure
              5.2d)
05         ((*Jaden lets his rubber band loosened.)) (Figure
           5.2e)
06  Gavin: [inside (??)]
           [((Jaden grabs his loosened rubber band with his hands
           and *stretches it toward the pin.))] (Figure 5.2f)
07  Martin:(exactly?) ((*Jaden stretches his rubber band over the
           pin. **Jaden loosens his rubber band to behind the
           pin.)) *(Figure 5.2g) **(Figure 5.2h)
08  Gavin: [IN:
09  Martin:[not [outside] outside will be
                [((Martin released his rubber band. Jaden presses
                down his rubber band onto the pin with his right
                fingers and stretches it with his left hand.))]
```

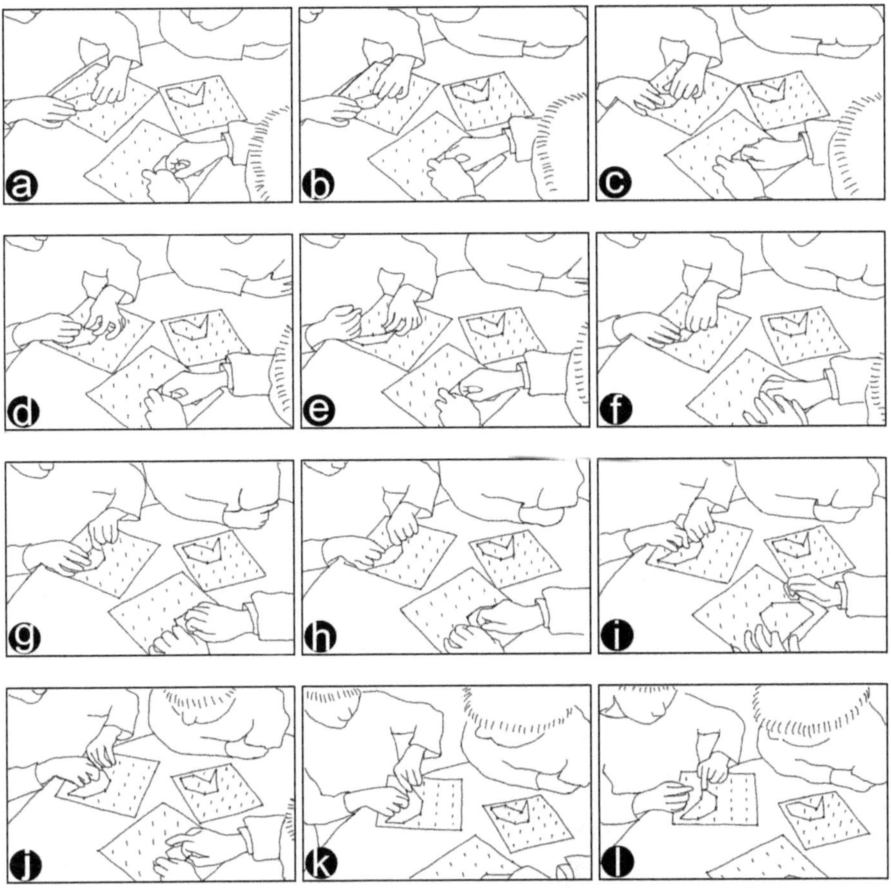

Figure 5.2. Jaden (left), Gavin (right up), and Martin (right down) sit at a table. a. Martin stretches his rubber hooked onto a pin and moves it back and forth behind the pin. b. Jaden holds his rubber band with his left hand and changes its position from one side to the other of the pin. Gavin turns his face toward Jaden. c. Jaden holds his rubber band with his left hand and grabs its right part with his right hand. d. Martin moves his rubber band back and forth five times behind the pin. Jaden stretches the right part of his rubber band around the pin. e. Jaden lets his rubber band loosened. f. Jaden grabs his loosened rubber band with his hands and stretches it toward the pin. g. Jaden stretches his rubber band over the pin. h. Jaden loosens his rubber band to behind the pin. i. Martin released his rubber band. Jaden presses down his rubber band onto the pin with his right fingers and begins to stretch it with his left hand. j. Martin grabs his loosened rubber band at the bottom and stretches it upward. k. Gavin changes his gaze toward Jaden. l. Jaden stretches his rubber band and hooks it to a pin on the left.

(Figure 5.2i)
10 [this (0.2) inside will be this]
[((Martin grabs his loosened rubber band at the bottom and *stretches it upward.))] (Figure 5.2j)
11 Gavin: ((*Gavin turns his gaze toward Jaden.)) (Figure 5.2k)
12 [ya:]
[((Jaden stretches his rubber band and *hooks it to a pin on the left))] (Figure 5.2l)

Description

The three boys sit at a desk in a mathematics classroom (Figure 5.1). On the desk, there is a sample shape (a polygon) that Gavin has produced (Figure 5.2a). Jaden makes a shape with a rubber band on his board. Martin holds a part of his rubber band on his board and says "remember, look at" (turn 01, Figure 5.2a). He utters "inside" in increased speech intensity and simultaneously stretches his rubber band back and forth behind a pin (turn 02, Figure 5.2a). Martin continues by uttering "down the" and Jaden simultaneously changes the position of his rubber band from one side to the other side of a pin (turn 02, Figure 5.2b). Gavin repeats the same word "inside," but he does so with increased speech intensity (turn 03). Jaden keeps holding the rubber band with his left hand and grabs the right part with his right hand (turn 03, Figure 5.2c). Martin repeats "inside" twice more and moves his rubber band back and forth five times behind the pin (turn 04, Figure 5.2d). Jaden stretches the right part of his rubber band around the pin (turn 04, Figure 5.2d) and lets it loosened (turn 05, Figure 5.2e). Gavin repeats "inside" and Jaden grabs the loosened rubber band (turn 06, Figure 5.2f). He stretches it toward the pin below (turn 06, Figure 5.2f). Martin says "exactly" (turn 07). Jaden stretches his rubber band over the pin and returns it back to a position behind the pin (turn 07, Figures 5.2g, h). Gavin speaks with greater speech intensity (turn 08). Simultaneously Martin says "not outside" (turn 09). He releases the rubber band, and thereby lets it unhooked away from the pin (turn 09, Figure 5.2i). Jaden presses down the rubber band onto the pin with his right fingers and stretches its left part with his left hand (turn 09, Figure 5.2i). Martin continues saying that "outside will be this" (turn 09). Martin grabs the loosened rubber band and begins stretching (turn 10, Figure 5.2j). He utters "inside will be this" and hooks his rubber band on the upper pin (turn 10). Gavin gazes at Jaden (turn 11, Figure 5.2k) and says "ya" (turn 12). Jaden hooks the left part of his rubber band onto a pin beside (turn 12, Figure 5.2l).

Analysis

In this situation, Martin holds the rubber band with his hands and says "remember, look at" (turn 01). Martin's speech and his body orientation constitute a conversational turn that calls for the other two boys' attention. Martin continues by saying "inside down the," which overlaps with his back-and-forth hand movement with the rubber band around the pin (turn 02). Simultaneously, Jaden changes the position of the rubber band with his left hand from one to the other side of the pin behind (turn 02, Figures 5.2a, b). Therefore, Martin's actions constitute communicative resources for concretely articulating structural conditions that enable a corner that is bent toward the inside of a polygon on his board (i.e., the same shape with the sample). The current position of the rubber band on Jaden's board (Figure 5.2a) does not allow Jaden to make the same shape of the corner on his board. Once the left part of the rubber band is stretched down, the rubber band would slip. Martin concretely shows this condition by saying "inside" with increased speech intensity and by moving his rubber band back and forth on one side of the pin. Those bodily actions thereby highlight where his rubber band is located around the pin. Martin's actions constitute the sense-making resources required for distinguishing the sample shape on Gavin's board from Jaden's shape. A way of talking geometrical inscriptions emerges in the presence of them. Therefore, the example shows how knowledgeability is articulated by means of the corporeally produced communicative resources including verbally spoken words, speech intensity and pitch, and hand movements. The living body constitutes the source of cultural means for signifying a physical structure that is materially present at hand but not yet salient (invisible) for Jaden.

The increase of sense-making resources in communication increases possibilities for the further development of a network of cultural significations. Martin continues talking (turn 02), which overlaps Gavin's utterance "inside" (turn 03). Martin utters "inside" two more times and moves his hands holding the part of the rubber band five more times (turn 04, Figure 5.2d). The repetition of literally the same word (with different prosodies) together with wider hand movements than before constitute sense-making resources for presenting the position of the rubber band even more saliently. He then provides another set of sense-making resources. Martin releases the rubber band and leaves it at the bottom of his board (Figure 5.2i). He utters "not outside, outside will be this," which constitutes a contrast to his next utterance ("inside will be this") and hand-movement that again hooks the rubber band up to the pin (turn 09). He produces additional sense-making resources (the word "outside," hand movements) that further clarify the spatial configuration of the polygon and the position of the rubber band. This increases the resources available for communication participants. These resources – words, prosody, and corporeal actions – are different but nevertheless are manifestations of the same communicative unit. This means that there is an increase in the heterogeneity of this unit. This increase of heterogeneity develops a form of communication about the two mathematical inscriptions in which each manifestation marks sense with respect to the others. For example, the word "outside" stands together with "inside" and the pair of terms stands for the position of the pin and the spaces separated by the rubber band.

The development of cultural signification increases the possibilities for knowing available to Jaden. He changes the position of the rubber band from one side to the other when Martin talks (turn 02). Then he presses down the rubber band to the pin and stretches the right part to make a sharp corner (turns 03–04, Figures 5.2c, d). Yet, this action turns out to be unsuccessful as it leads him to turn the rubber band around the pin almost to the other side (180°). Jaden releases the rubber band and moves his hands away from it (turn 05, Figure 5.2e). Jaden grabs the loose rubber band with his hands and stretches it again down to the pin (turn 06, Figure 5.2f). Gavin's utterance "inside" (turn 06) overlaps it. Jaden continues stretching the rubber band, which actually passes over the pin (turn 07, Figure 5.2g). Jaden loosens the rubber band and places it on the other side of the pin (Figure 5.2h). Jaden presses the stretched rubber band down with his right hand (turn 09, Figure 5.2i). This allows him to stretch the left part of the rubber band with his left hand without having to change the angle of stretching. The series of coordinated action exhibits an increasing competency in dealing with the present structural conditions, which allows him to hook the left part of the rubber band to a pin on the left (turns 10–12, Figures 5.2j, l). Together, the board, pins, and rubber band constitute a part of the world that Gavin's finger navigate with increasing facility – the fingers, and therefore the "I can," increasingly know their way around this world. Geometrically knowledgeable actions appear as Jaden participates in communication (listening, seeing) while he actively engages in the deployment of resources. The totality of the different communicative resources influences this organization. Jaden's action appears together with the unfolding of Martin's talk. The two are concurrent rather than in a temporally linear cause-effect relationship. Here, knowing mathematical inscriptions emerges from the different, irreducible modes of communication as an integrated whole.

Discussion

The communicative action in this reproduction task exemplifies the real-time development of participative (unindifferent) understanding in which an arbitrary shape is juxtaposed with the children's cultural and linguistic resources. With respect to these, the shape comes to stand for something – a geometrical inscription that can be reproduced and shared within this group or with others, for example, when they report back to the class. A child may reproduce a shape on a piece of paper by using a pencil, by using a computer, or by using materials such as wooden sticks. The case shows that the pedagogical practice for assisting someone in reproducing the shape is built upon a series of culturally signifying moves and that this work is done in and through the living body, which makes, for example, a distinction between inside and outside around a pin. Martin's word "inside" marks sense by being juxtaposed with his holding a part of rubber band on one side of a pin and with his talking while gazing at the position of Jaden's a rubber band. Martin's talk, intended to assist his peer, might have increased the amount of sense making work that Jaden has to do, because now he has to attend to both the original sample and Martin's talk over his board and make a connection between them. In a Vygotskian framework the potentially endless process of signification would stop only when communication and thinking stop. These are two lines of development that come to be related as Jaden thinks with his rubber band and board while he is hearing and gazing at the other talking. At this point of the encounter, the senses of the words "inside" and "outside" are coordinated with the material conditions of the board. Therefore, the role of the pedagogical discourse is to let this encounter between communication and thinking happen. The encounter is made possible in and through the living body where seeing, hearing, feeling the shape, and modifying the rubber position come together and come under the power of the "I can." The following three assertions summarize the three aspects of the sense making process: (a) the living body reproduces and transforms cultural resources and connects them to mathematical inscriptions; (b) the increase of heterogeneous resources in communication increases possibilities for the development of a web of signification; and (c) participative (unindifferent) understanding of mathematical inscriptions emerges from the combination of the different modes of communication that form a unified but heterogeneous whole.

TALKING INSCRIPTIONS AND THE INSCRIPTION OF CULTURE

> How does the geometrical ideality (like that of all sciences and similar ideal forms) get from its intra-personal originary source, where it is a form in the space of consciousness of the inventor's soul, to its linguistic embodiment? (Husserl, 1939, p. 210)

Husserl asks how anything that was invented somehow comes to be embodied in language, and, thereby comes to be shared among the members of a culture. He answers his own rhetorical question by saying that the very possibility of speaking a thought makes any namable inner form immediately an external, public form. This is why Vygotsky speaks about the coincident development of speech and thought. Precisely because knowing was produced by and leaves a trace in the living body, can it be reproduced not only by the same living body but, because the living body is a concrete form of life, by all living bodies of the same structure. It is precisely when the original subject can recall what it has thought that this remembered form may also be reproduced and recalled by others. This is so because the very nature of recall requires the capacity to make something present again, that is, to work with *re*presentations. *Re*presenting means *re*producing, and reproducing never is mere repeating. This also means that there is evidence of the same, for that which is *re*produced is "identical" to the original production. "Together with the originary production comes the possibility of arbitrary number of identical repetitions of the form in a chain of repetitions" (Husserl, 1939, p. 211)

Participative (unindifferent) understanding of mathematical inscriptions is evidenced in social practice where heterogeneous communicative resources mediate children's experiences. *Social* practice inherently means repeated in "identical" form by different individuals, that is, sufficiently close for the repetition to be part of the same family. It is therefore through their subjective sensual expressions that children reproduce the objective nature of the formal disciplines. This is so because "sensual expressions obtain spatiotemporal realization as all corporeal phenomena, as anything that is bodily embodied. Not spatiotemporally realized is the ideal form itself, the 'ideal objectivity,' the 'sense' of the sensual expression and documentation" (Husserl, 1939, p. 210). Culture is embodied in objects (bodies), which, when they are used, give shape to children's actions making these inherently cultural.

Culture, inscribed into the objects (bodies), comes to be inscribed in the bodies of the children who use them. In this chapter, we articulate the inherently cultural dynamics of sense making from a holistic perspective. We exemplify the fact that the living body constitutes a place and means for making communication resources available and for developing webs of signification. It is in the body that living labor comes to be experienced and recorded. It is in and by means of the living body that talk comes to be linked up with the experience of inscriptions. It thereby constitutes something like a hub where and from which develop the dynamic relation between communication and mathematical inscriptions. The three assertions presented in this chapter propose that this encounter appears from the totality of communication performances. Mathematical inscriptions come to stand for something not by its own structure but in relation to a web of signification. Therefore, knowing mathematical inscriptions is equivalent to reading between different sense making resources – as if we were reading between different lines – to bring about the sense of a mathematical inscription in its totality. The living body performs this reading by hosting cultural significations in the inscription-related communication.

Initially, the new understandings are passively received as new relations come to be suddenly discovered (Husserl, 1939). But because "everything has a name and is namable," "the psychic as psychic of this human being is eo ipso 'objective,' that is, in its re-understanding made accessible intersubjectively" (p. 211). That is, these children reproduce the objective nature of science and mathematics precisely because each shares in the experience of the living labor even though the content of this experience may differ between them. The very possibility for each child to remember what it has done and said lies at the origin of the reproducibility of the science: "past experience may be lived again and actively" (p. 211).

Competent use of mathematical inscriptions is essential to the notion of scientific and mathematical literacy. Mathematical inscriptions are treated as the material body of scientific and mathematical thinking and therefore the (pure) vocabulary of literacy. It is in fact only through these inscriptions that mathematics historically could become the objective science that it is (Husserl, 1939). This chapter shows the *contingent* nature of mathematical inscriptions, which means (a) that mathematical inscriptions do not stand by themselves but are co-present with other bodies within webs of cultural signification and therefore (b) that mathematical inscriptions never constitute a self-explicating medium of mathematical thinking. Perhaps this is why teachers still have to spend a lot of time preparing mathematical inscriptions for their classroom talk despite a number of good inscriptions that curriculum developers supply. Mathematical inscriptions constitute the material body that makes the objective nature of mathematics possible. The three boys are enabled to communicate a way of looking at the polygonal shape because of its objective presence.

CHAPTER 6

HETEROGENEOUS PERFORMANCES AND LINGUISTIC HYBRIDITY

> But without the really formed capacity of a reactivation of the originary activities that are embodied in the fundamental concepts and of its prescientific materials, geometry would be a tradition without sense. A tradition of which, if we had lost this capacity, we could not even know whether it has or ever had a sense. (Husserl, 1939, p. 216)

The philosopher reflecting upon the origin not only of geometry but also of all sciences notes that there are originary, prescientific experiences that are build into the scientific concepts. In reactivating these concepts, we also reactivate the prescientific "materials" that have led to the first, originary formulation of the science. That is, the nonscientific language is the material and tool that has led – and as we maintain, in the case of children and students of today still leads – to the formulation of scientific concepts. Rather than having to be abandoned or, as some scholars say, eradicated, the originary forms of (informal) language and concepts that children bring to schools are actively exploited in the formation of scientific (formal) language and concepts.

In this chapter, we articulate an approach to science talk that does not start from the dichotomy of the formal and the informal genres of language. We theorize the sense-making role of science talk by taking a cultural-historical and incarnate approach to language. This approach considers the different, irreducible modes of communication – sound-words, (hand, arm, head) gestures, body position, body orientation, prosody – as an integrated whole and therefore allows theorizing scientific and mathematical thinking without dichotomizing it into concrete versus abstract. We provide an exemplary analysis of science talk in a university physics lesson and exemplify our dialectical theory that explains the development of conceptual understanding along with the development of everyday language. We suggest that this approach shows how people communicate scientific concepts through the hybridization of communicative means. We conclude that the continuity of conceptual understanding involves dis/continuity.

BEYOND THE DICHOTOMY OF LANGUAGE

Language takes a central place in theories of learning science because knowing constitutes a way of talking. Learning to talk constitutes a self-reference, for talking mediates talking. That is, based on existing forms of talking, which constitute both topic and ground, we talk up new forms of language (Roth, 2010b). Language is an important entry point into understanding learning and the handing down of science because it "creates the possibility of a human community of sympathetic understanding and its correlate – the common world of objects" (Husserl, 1939, p. 210). Between these two related aspects of language, scholars theorize science and mathematics as a special kind of culture distinguished from others.

Some scholars think learning as an accommodation to the special language (i.e., genre). Other researchers attend to students' everyday linguistic interactions and understand that everyday talk is the condition for rather than a hindrance to learning to talk science and mathematics. The simultaneous consideration of the two different aspects of language – informal versus formal – constitutes an inner contradiction even in the case where educators want students to abandon or eradicate their vernacular talk. This is so because the abandonment and eradication requires existing language as a support and tool. But the new and different way of talking, which students are supposed to learn (i.e., science), constitutes the necessary condition for achieving conceptual understanding: This new way of talking *constitutes* the conceptual understanding. From a cultural-historical perspective, a suitable way to resolve the dialectical problem of language is to abandon the dichotomy of the legitimate (formal) and the illegitimate (informal) and to acknowledge students' participation – however unscientific and prescientific this may be – as an instance of the cultural practice of talking science. That is, the very existence of a trajectory from informal to formal talk requires hybrid forms. These hybrid forms of talk are integral part of the culture of science because they characterize the trajectories of becoming a scientist.

In this chapter, we also acknowledge other work on learning that does not dichotomize formal and informal language use (e.g., Lave, 1997). We take a cultural-historical and dialectical perspective and articulate concepts for holistically understanding language. We suggest that the analysis of literacy for concept development needs to consider the non-self-identical nature of language. We show that the concepts of hybridization and boundary crossing allow us to understand how learning to talk scientifically is possible even when we acknowledge the prescientific and unscientific. We provide concrete analyses of case examples selected from an undergraduate physics lecture.

HETEROGENEOUS LANGUAGE AND CONCEPT DEVELOPMENT

Teaching inherently presupposes that what children and students know is insufficient. More so, the teaching of science presupposes that the everyday understandings of the world that students bring with them to school are inappropriate and therefore are in need of change. Students often experience the sciences in the way that colonized people experience the confrontation with another religion, culture, or language that they are made to acquire during the colonization process. The very fact of schooling produces discontinuity, which Derrida (1996) describes when he, as an Algerian Jew, encountered French literature in the school he attended. Thus, there was a discontinuity that

> became, in the fact, *doubly* revealing. It exhibited without doubt the heights that always separates the literary culture – the "literarity" as a certain treatment of language, of *sense* and reference – from the non-literary culture, even if this separation never reduces itself to the "pure and simple." But other than this essential heterogeneity, other than this universal hierarchy, a brutal severance was, in this particular case, fostering a more acute partition: the one that separates French literature – its history, its works, its models, its cult of the dead, its modes of transmission and celebration, its "posh districts," its names of authors and editors – from the culture "proper to "French Algerians." (p. 77)

To complete the analogy, all we are required to do is take science in the place of French "high" literature and oppose it to the mundane experiences students bring to school, which, in the analogy, take the place of the culture of the French Algerians.

The division of literacy into the formal and the informal is one of the dualisms that can be observed in many science education studies on conceptions, conceptual development, and conceptual change – regardless of whether the informal is described positively (e.g., informal science literacy) or negatively (e.g., naïve ideas). The very acknowledgment of a dichotomy leads to theorize science language as a self-identical entity. This entity stands in opposition to the vernacular language. How students can ever move from the vernacular to the formal comes to be mystified or the problem is simplified by saying that students learn the new language and then go back and forth by switch codes. From a cultural-historical perspective, a minimum unit required for analyzing knowing and learning is a societal activity (e.g., doing baseball, schooling, or doing research on learning science) in and for which participants communicate (e.g., Leont'ev, 1978). On the surface, thinking in terms of different activity systems leads us to think in terms of different languages: In an activity system of science, the language is scientific and formal; in an everyday context, the language is vernacular and informal. Once approaching the learning problem in this manner, scholars see themselves forced to talk about the boundaries between different activities and the need for learning how to cross the boundaries or how and when to switch from one code (e.g., the formal) into another code (e.g., the informal). Other scholars use the concept of third space to model how students somehow hybridize language and culture from the two activity systems. But, we have to point out, students are not just confronted with two ideal activity systems: Their biographies are interwoven with cultural history. Becoming as a person, that is, developing a personality means evolving a network of experiences in which the object/motives of the activity systems, and therefore the mutual relations of language and culture, come to be integrated (Roth, in press).

Our model of language leads us to a different understanding of literacy than that of the code-switching model. We take a dialectical perspective in which "development" (in learning to use a new language and culture) is not the homogenization of a code but an increase in heterogeneity of that code. Becoming literate, knowing one's way around the world, is equivalent to an increase of communicative means. Increase in literacy means translating cultural possibilities that are always already

Figure 6.1. A physics professor talks about thermodynamics concepts to third-grade physics undergraduate students at a university lecture hall while producing a graph.

hybrid rather than pure forms (Nancy, 1993a). Thus, it is important to think literacy in terms of an increasing hybridization of language-in-use. This also constitutes a way of taking a non-reductionist approach to the dynamic system of participative (unindifferent) understanding, because any word-in-use comes with heterogeneity because it bridges speaker and hearer.

We exemplify a theoretical framework that addresses the heterogeneous nature of literacy and culture and their significance to theorizing conceptual understanding. We develop a cultural-historical and dialectical approach to language for theorizing conceptual understanding without dichotomizing formal versus informal or cognition versus language. Following other studies, we integrate speech and other heterogeneous communicative performances into one unit (Roth & Pozzer-Ardenghi, 2006). That is, the communication of scientific concepts is understood in terms of the complex performance of my living body that produces different forms of expressions simultaneously. These different forms, modalities, are not due to a single cognitive act behind them – I do not think what I am saying in everyday conversation ahead of saying it; my saying is my thinking. Nor can they be understood to be the result of independent acts – it is a single (rather than split, multiple) me who expresses *a* thought even if the expressions are very different. The different modalities have to be understood as manifestations of a higher unit, which we, following phenomenology (e.g., Henry, 2003) understand to arise from the singular "I can" of my living body. We take a 46-second-long example of university physics lecture (one of the typical contexts for talking science) in which a physics professor explains a thermodynamics graph and associated concepts to his third-year university audience (Figure 6.1). We analyze the lecture talk and articulate the heterogeneous dynamics of it. We suggest that *hybridization* constitutes a productive concept for theorizing knowing and learning exhibited in such a lecture.

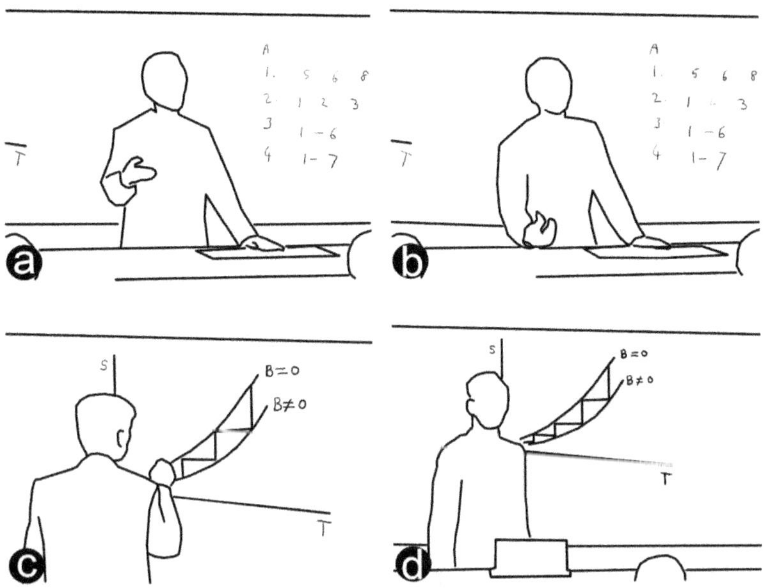

Figure 6.2. a. The professor holds up his right palm. b. The professor bounces his right hand on the table. c. The professor follows with a piece of chalk the stepwise lines between the two curves. d. The professor turns from the blackboard and faces the students.

This concept *sublates* – a term from dialectical theory that simultaneously denotes both 'to destroy' and 'to preserve' – the formal-informal dichotomy.

Transcript 6.1

```
01  [EVEN after this]
    [((*holds up his right palm and returns it back to the
    desk))] (Figure 6.2a)
02  bounced a million times, there are STILL a millions and
03  [still an INFINITE number of bounces]
    [((*bounces his right hand three times on the desk))] (Figure
    6.2b)
04  ahead of it (1.3)   [<<p>you've done this]    in high school
                        [((lifts his palms up))]
05  [or you're (??)>] (1.1) it is still infinite bounces ahead
    [((lifts his palms up slightly))]
06  of it, and: but it comes to rest after a finite time, after
07  having covered a finite distance (1.2)
08  and I CAN'T    [tell (one bring if?)]
                   [((gazes at his notes on the desk))]
09  one of you one of these days, maybe able to think of
10  [some process,]
    [((turns body to the blackboard, walks to the chalkboard))]
11  [by which you're]
    [((raises his right hand up and brings the chalk to the right
    end of the graph))]
```

12 [able, to move a]
[((**follows the stepwise lines of the graph on the board with the chalk*))] (Figure 6.2c)
13 [(0.6) an infinite number, of:, uh:,]
[((*moves his right hand away from the graph and gazes at the right top area of the graph*))]
14 [isothermal]
[((*follows the top vertical line*))]
15 [constraints (0.7) followed by uh]
[((*stops at the lower end of the vertical line and stay still*))]
16 [adiabatic]
[((*follows a top horizontal line from the left to the right*))]
17 [relaxation]
[((**turns his body from the blackboard to the students*))] (Figure 6.2d)
18 to infinite number of experiments like- like- in a finite amount of time

Description

The physics professor stands behind the desk and faces students with the blackboard (graph) at his back (Figure 6.1). The students sit toward the front side (the professor and the blackboard) of the lecture room. The professor utters "even after this" (line 01). He simultaneously holds up his right hand and returns it back onto the table; the body movement thereby produces an emphasis of the utterance (line 01). He continues by uttering "bounced a million time, there are still a millions and" (line 02). He describes a ball that falls and bounces up and down. He suggests that even after the ball bounces a number of times, there are still millions of times left to bounce. He says "still an infinite number of bounces" (line 03) ahead of it (line 04) and thereby points out that the number of bouncing is infinite. He lightly taps the desk three times with his right hand. This can be seen as an iconic gesture of a ball bouncing up and down on the top of the desk (line 03). The professor pauses for a second and then continues talking with considerably lower speech volume. He intimates that students have done this – the infinite number bounces – at high school, and thereby points out that he is talking about something that is (or ought to be) already known to students (lines 04–05). There is another 1-second pause (line 05). The professor repeatedly utters "infinite bounces ahead of it" (lines 05–06). He elongates "and" (line 06) and then states that this infinite number of bounces "comes to rest after a finite time" (line 06) and "after having covered a finite distance" (lines 06–07).

The professor declares "and I can't tell" and changes his gaze direction to his lecture notes on the table (line 08). He continues uttering "one of you one of these days, maybe able to think of some processes" (lines 09–10). He changes his body orientation from the students to the chalkboard and then walks toward it (line 10). He says "by which you're able to move a" and simultaneously raises his right hand and the chalk up to the right end of the graph and rapidly follows the step-forming lines of the graph from the right top to the zero point at the left down (lines 11–12,

Figure 6.2c). The professor now pauses and moves his hand away from the graph (line 13). He gazes at the right top of the graph and utters "an infinite number of" (line 13). He produces the interjection "uh," which can be heard as a hesitation (line 13), and then continues saying "isothermal" while his right hand simultaneously follows the top vertical line (line 14). The chalk stops at the lower point of the vertical line, following which the professor utters "constraints followed by" (line 15). This statement is followed by the utterance "adiabatic" and his body moves following the top horizontal line from the left to the right (line 16). He suggests that this adiabatic process pertains to "relaxation" and turns toward the students (line 17). The professor utters "to infinite number of experiments like, like in a finite amount of time" (line 18), and thereby highlights a thermodynamic process by which an infinite number of cycle isothermal constraints and adiabatic relaxation happen in a finite amount of time.

Analysis

In this situation, the professor explains a mechanical movement of a bouncing ball. He articulates that a bouncing ball requires an infinite number of bounces until it comes to a rest (lines 01–04). In classical mechanics, a ball dropped from a height that bounces with a coefficient of restitution lower than one reaches a new maximum height that is lower than the previous one. If the ball keeps bouncing at a constant rate (i.e., the coefficient is constant), the decreasing maximum heights constitute an infinite geometrical series. A bouncing ball always travels a certain distance although the maximum height becomes very, even infinitely small. The sum of a geometrical series is finite – i.e., a bouncing ball stops on a floor in a finite time after it covers a finite distance (for a worked out example with illustrations see http://www.sosmath.com/calculus/geoser/bounce/bounce.html). The professor reminds students of what they (were supposed to) have learned in high school (lines 04–05) and articulates this principle of motion (lines 05–07). His suggestion that they have done "this" in high school (line 04) exhibits his presupposition that students are already familiar with this kind of motion.

The professor's talk about a bouncing ball constitutes a form of hybridization. One may claim that the professor is talking about a topic that does not constitute the genre that he is supposed to speak in this lecture course. The physics of a bouncing ball may not belong to thermodynamics at all but to another topic of physics (i.e., classical mechanics) or to some everyday event that the students are familiar with from playing with balls. People easily find examples of bouncing balls in their everyday lives. Whereas thermodynamic cycles – such as the adiabatic process that the professor currently is talking about – usually involve complicated machines that implement cooling, heating, and increases or decreases in pressure, a ball and its bouncing motion constitutes an everyday phenomenon that also is a topic in courses on classical mechanics. Here the professor realizes the topic as a constitutive part of his thermodynamic lesson. He brings together two very different phenomena and forms of talk, that is, he hybridizes talk, and thereby brings about an analogy. That is, when an analogy draws on an everyday phenomenon and on the lived experiences of students, the talk hybridizes formal and informal expressions and experiences.

The talk taken together with other embodied communicative productions (e.g., gesture, prosodies of speech, and body position) produces a specific (hybridized) genre that addresses what students need to understand and learn to speak as part of their participation in this lecture talk. For example, the pitch of his speech moves higher when he utters words "even" (line 01), "still" (line 02), and "infinite" (line 03), which thereby highlights that there are a number of upcoming bounces and it is actually infinite. The professor's articulation "theorizes" the motion of a bouncing ball using a particular framework (i.e., infinite bounces, finite time, and finite distance). The professor translates into his own performances the physical motion of a bouncing ball and produces expressions for making sense of the phenomena; these means do not draw on verbal consciousness but on different forms of awareness, here for prosody. These expressions constitute hybridized resources that anyone in the lecture hall may use to continue talking in the next. The presence of the audience (i.e., undergraduate students) and the professor's orientation toward them conditions this hybridized form of lecture talk. For example, the professor stands toward the students and use a podium as part of his speech (e.g., bouncing on the table). Moreover, his mention of "high school" and his expressed assumption (lines 04–05) explicitly shows that his talk is *oriented to* and thereby *designed for* a specific audience. That is, the expression cannot be attributed solely to him but rather constitutes a characteristic of the audience as well – just as this was articulated in a dialectical, Marxist theory of language (Bakhtine [Volochinov], 1977). Therefore, the professor hybridizes thermodynamics talk and content (i.e., the graph on the board, Figure 6.1) by infusing into it vernacular talk about a bouncing ball.

Following the topic of the bouncing ball, the professor returns to the analysis of the thermodynamic cycle represented in the graphic on the board (Figure 6.2). The professor appears to be searching for ways to continue when he gazes at his lecture notes (line 08). The professor then articulates the possibility of a thermodynamic process that includes an infinite number of isothermal-adiabatic cycles within a finite amount of time (lines 12–18). Simultaneously, the professor's hand movements (e.g., following the lines [lines 12–16]) enact a stepwise decreasing motion and thereby realize a physically similar pattern of motion that he has already produced together with the talk about the bouncing ball. The professor does not explicitly say that the bouncing ball functions as an analogy for the isothermal-adiabatic thermodynamic cycle. Yet, his body position and orientation, his hand movements, and finally his utterances of "infinite number of experiments" (line 18) and "a finite amount of time" (line 18) offer up opportunities for perceiving the two situations as analogous. Moreover, why would the professor talk about a bouncing ball unless it had some bearing on the present situation, which is the talk about a thermodynamic cycle in a course on thermodynamics? Thus, the corporeal performances not only articulate the thermodynamic cycle in a structurally similar way with the motion of a bouncing ball but also explicitly let the latter work as an analogy for the former. This means that the same expressions function to explain two very different situations – the expressions not only are heterogeneous *within* themselves, consisting of very different modalities, but also *between* themselves referring to and denoting very different physical phenomena.

From a dialectical perspective, the professor's embodied performances constitute a heterogeneous unit that cannot be reduced to a simple combination of two different topics (i.e., a bouncing ball and a thermodynamic cycle consisted of isothermal constraint and adiabatic relaxation). This is so because the embodied performances not only produce the structural similarities but also constitute a hybridized form of talking. The professor talks about making an infinite number of cycles that consist of isothermal constraints and adiabatic relaxations within a finite time. Not only does he use the same words (i.e., "infinite number" [lines 13, 18], "finite amount of time" [line 18]) but also he employs the same hand movements temporally coordinated with the words (e.g., turning toward the graph [line 10], following the stepwise lines between the curves [line 12, lines 14–16], gazing at the graph [line 13]). These performances constitute resources for taking the case of a bouncing ball motion as an analogy for understanding the physical properties of the thermodynamic cycle; but simultaneously, the thermodynamic cycle can be taken as an analogy for the bouncing ball because of the equivalence in the analogical relation. The heterogeneity of linguistic performance constitutes hybridity that transcends the purity of linguistic expression, on the one hand, and the purity of signifieds, on the other hand. But it is precisely this hybridity that offers opportunities for overcoming the inherent contradiction that exists when we think about how students, who do not know physics and physics language can be "introduced" to this subject matter and its discursive forms.

HYBRIDITY AND HETEROGENEITY

In the preceding section, we present an example of a physics lecture highlighting the heterogeneity of language that *has to be used* to bridge the gap – if there ever existed one – between purely scientific code and a purely vernacular code. But such talk about different codes leads us into the wrong direction when we attempt to understand learning. This is so because on both cultural-historical and individual developmental grounds, the everyday language (code) not only always already precedes the formal code but also provides the very means to elaborate and achieve a formal linguistic code. Husserl (1939) therefore suggests that any question about the origin of science has to be developed based on the "knowledge of the principal structures such as 'originary production' [Urstiftung], 'originary evidence' [Urevidenz,' 'originary material' [Urmaterial,' sedimentation, reactivation, and so on" (p. 221). In the preceding analysis of the lecture episode, we show that the increase of the heterogeneity (hybridization) and the transaction between different corporeal performances constitutes a central aspect of the continuity that the professor's lecture talk realizes between the motion of a bouncing ball and the movement of a thermodynamic system. The professor introduces students the concept of the third law of thermodynamics by proliferating the heterogeneity of communicative resources and by offering possibilities for participative (unindifferent) understanding to arise in his relation with the student audience. Developing a better understanding of the thermodynamic cycle (the graph) requires a better understanding of the motion of a bouncing ball in this case. The two understandings develop together as one hybridized communicative unit because of the bi-directional nature of the analogical relation. In this framework, we do not separate conceptual understanding into cognitive and linguistic events because of the interrelated and – according to Vygotsky (1934/1986) – the mutually constitutive nature of the two processes. Thinking and communicative performance are but two one-sided manifestations (expressions) of a holistic dynamics that we denote here as participative (unindifferent) understanding. Learning to talk science implies a trajectory of hybridized genres that have both vernacular and scientific aspects. This is structurally equivalent to the situation where I – as we both know from our auto/biographical experience – learn a new language different from my mother tongue taking the latter as the tool and ground for acquiring the former. What we need to better understand is how the scientific genre eventually can come to stand apparently on its own – well knowing that this is only an appearance, because all scientific language presupposes and is build on the vernacular.

In this chapter, we propose approaching learning from a cultural-historical perspective that does not dichotomize everyday vernacular and scientific language. In this approach, communicative hybridization and heterogeneity are central ideas. We show that learning presupposes everyday understanding and the vernacular, which serve as the material and means for new linguistic genres (here the scientific) to emerge. To talk about and make sense of new experiences, students have no other recourse than to draw on what they are already familiar with. That is, even in the case that the vernacular is inconsistent with the scientific language, learning must draw on the former to attain the latter. That is, the hybridization is central to the continuity of individual and cultural development. This is consistent with the approach Husserl (1939) asks us to take when he focuses on the originary sense that constitutes the foundation and source of science and that is continuously reproduced together with the science. That is, even when students eventually become competent users of a scientific language, the original language and sense continues to be present. What Husserl says about geometry is valid, as he suggests, for all sciences: "From the perspective of what we know, from our geometry as a tradition of knowing, it is possible in principle to ask a question about the *sunken* originary beginnings of geometry in the way these must have existed necessarily" (p. 207). Thus, the originary even pre-linguistic experience with bouncing balls is part of the origin upon which subsequent conceptual talk about the event can be founded in terms of the logic of geometrical series.

PART C

THE BODY IN A STRANGE WORLD

> The question of the mode of experience, in which the foreign/strange enters, cannot be separated from the larger question of how we come to encounter this foreign/strange. To this question corresponds the motive of responsivity, which cannot be thought without a touch of ethics. (Waldenfels, 2006, p. 56)

Learning something new, whether I am a child in a second-grade mathematics course, a university student in a third-year physics course, or a researcher investigating knowing and learning in the two settings, means confronting the foreign/strange. For if I were already familiar with the curricular topic, there would be no need learning it. Traditional theories conceptualize learning from the perspective of the knower, that is, from the perspective of someone standing outside the learning process like a researcher watching rats running a maze. But from my perspective as the learner, the process has to look different, because I do not know what I am to learn. This mode of experience, therefore, is different than what traditional learning theories presuppose it to be. More importantly, considering learning from the perspective of the unknown immediately leads us into ethical questions. These two moments of the learning experience – the unknown and the ethical question – constitute the topic of this third part of our book.

Scientific and mathematical sense making sometimes is treated as a problem of learning to use some specific language and making relevant propositions. Yet, what does it mean to be able to produce a proposition by using a specific language and its system of categories? The preceding sections of this book exhibit our approach to conceptual understanding, which differs from the more common mainstream approaches that attempt to derive explanations focusing on the contents of mind as if such could exist without the living body. A category marks sense through a series of translations, which, in our approach, are understood as a corporeal performance (e.g., the concept of "sphere" in chapter 1). Following this approach, we show that sense making and higher-order cognitive functions are inseparable from and integral to practical activity, the embodied experience of culture that inherently has a history. In any human activity, objects provide opportunities for concretely realizing cultural possibilities and thereby mediate the corporeal experience of culture as part of praxis (e.g., see chapter 2). Learning is the result of the hybridization of prior experiences and novel forms of language (e.g., chapter 6). The relevant unit for understanding the process of sense making and the associated, generative aspects of learning (e.g., conceptual development) does not lie in a specific form of representation but in the communicative performance as a whole. This whole cannot be expressed by verbal means alone, for the latter is only one aspect of the communicative totality. Sense making consists of and exists in the labor associated with communicative performances. This raises questions: "How can we approach sense making in science and mathematics such that the development is related to performance and learning?" "How can we theorize the historical development of scientific and mathematical sense making without privileging some representations?" and "How do people experience the change of the world that they inhabit and their sense making in mathematics and science?"

In this part C, we theorize sense making from an anthropological perspective. We stipulate that the expansion of the human collective ground for action (i.e., solidarity) constitutes the condition for successful sense making (e.g., critical psychology). That is, the living body mediates sense making because it allows for solidarity. We show that sense making consists of the bodily translation of culture experience such that it expands the possibilities for acting in the world. This expansion occurs in the direction of that which is previously unknown, which requires us to think about the role of the unknown, the foreign/strange, both as a mode of experience and as a way of approaching a theory of learning. The significance of solidarity lies in the possibility for articulating theories of knowing and learning that are reflexive of the lived experience of culture and therefore open to new, previously foreign/strange forms of performance that arise from it. We understand that this approach to sense making offers a better grasp of the process by which a human collective life (i.e., culture) is reflected in the individual human psyche (i.e., mind), and constitutes a response to Vygotsky's concept of a *zone of proximal development*. Learning science and mathematics is comparable to learning a foreign language and therefore requires participation in *collective* practice of speaking that language. Yet, making sense of a new language means dealing with the plurality not only of the known and foreign languages, but with the plurality of language as such. We need to consider that "this other ontological region in which something like a body can appear and develop is a heterogeneous and essentially different reality" (Henry, 1975, p. 3).

The two chapters in this section provide an analytic framework of communication that integrates emotional-volitional valence and ethico-moral valence of action at the heart of cognition. In chapter 7 we portray the living body as the phenomenon that allows us to deal with the variegated nature of our social and material realities. Thus, the emotional-volitional dimension constitutes an important aspect for understanding the formation of growth points of new ideas, that is, a configuration in which all communicative resources are coordinated in such a way to expand *room to maneuver*. In chapter 8 we present the living body as the ground upon which solidarity and responsibility for the other emerge. We show that generative aspects of scientific and mathematical sense making arise from relevant performative, situated, and distributed dimensions.

CHAPTER 7

LEARNING – FROM THE PERSPECTIVE OF THE UNKNOWN

> How can we consider the foreign/strange without already neutralizing or even denying, in the way we are dealing with it, its effects, its challenges, and its aspirations? (Waldenfels, 2006, p. 9)

To learn means to come to know something heretofore unknown. While still unknown, we inherently cannot aim at an object (e.g., knowledge, concept). We cannot even know that there is something like an object to be parceled out. The unknown is both massive and undifferentiated and known only through its presence as non-presence. But how then can we *intentionally* engage in learning something we do not know if, by its very nature, we cannot aim at this unknown object, especially as we cannot know of its existence within the massive, undifferentiated whole of the unknown? We do not even know that there is an object of a particular kind. How can we think learning from this perspective without, as our introductory quote suggests, already neutralizing or even denying the very nature of the situation? What can and do we have to do to theorize this aporic situation whereby intentional learning of an unknown (epistemic) object is inherently impossible?

In this chapter, we propose a set of concepts for thinking about issues of learning science. We take the perspective of the unknown, that is, learning as an encounter of the learner with the (radically) foreign/strange. Because the foreign/strange is unknown, we cannot see or envision it. We cannot aim at – the French have the verb *viser*, to aim at, which has the same stem as vision – or intend it. What understand that which we eventually come to see as arising from the invisible, unseen, and therefore unforeseen (Marion, 2004). We draw on ethnographic work in one undergraduate physics course at a Canadian university, where we followed one female Japanese student in particular through different activities in her life as a student. She had come to the country for the purpose of getting a degree. Leaving one's country and entering a new culture constitutes a mode of experience that provides us with some leverage to understanding learning more generally. This is so because "remoteness, absence, distance, even instants of solitude and dislocation, on all which phenomenologists focus on in their analysis of the experience of the foreign/strange, do not constitute impairments of this experience, but belong to its essence" (Waldenfels, 2006, pp. 56–57). As an entry point and as source of empirical materials, we draw on our own auto/ethnographic experience that brings particular advantages to ally pathos to the experience of the foreign/strange, something is happening to (affecting) us that is beyond all experience, understanding, and anticipation. We articulate three aspects that pathos (empathy) allows us to understand concerning the experience of the foreign/strange and then provide an exemplary and exemplifying analysis.

OTHERNESS REVEALS THE FOREIGN/STRANGE

> Other living-bodies [*Leiber*] are bodies [*Körper*] in rest and motion (always: forward-motion [*Fort-Bewegung*], in the sense of approaching or moving away), but there are living-bodies in the "I move," whereby the I is "other I," for which my living-body [*Leib*] is a body [*Körper*] and for which all external bodies [*Aussenkörper*], which are not living-bodies [*Leiber*] for it, are the same [external bodies] that they are for me. But even every living-body [*Leib*], which is foreign living-body, is an "I" for all others . . . (Husserl, 1940, p. 315)

In this opening quote, Husserl makes a distinction between living bodies and non-living bodies. He notes that other living bodies are constituted as such, even though they are, in one sense, also just material, non-living bodies. Thus, even though another living body constitutes something foreign/strange, it is also familiar in the sense that it appears to me as a living one. This relation between living bodies constitutes the foundation of empathy, even when I encounter someone from a culture other than my own. The possibility for empathy is heightened, as the previous quote from Waldenfels shows, through our own experience of remoteness, absence, distance, solitude, and dislocation. In this era of rampant globalization, the encounter with individuals from foreign/strange cultures is becoming an increasingly common experience.

Globalization has become a catchword in many scholarly disciplines, but originally had been developed and used in economics. Derived from the adjective global, which characterizes the pertinence of something to a totality of items, the term is used to think issues that pertain to the whole world. For example, information and monetary exchanges now involve trade networks that reach to the farthest corners (countries) of the Earth. With the accessibility of and to informational media, such as Internet radio and television, diverse cultures are confronted with the phenomenon of globalization, where they, already heterogeneous at heart, come to encounter the unspeakable and unthinkable other. Science education, too, is confronted with the phenomenon (experience) and concept given the fact that international migration – because of voluntary emigration and immigration as well as involuntary displacement and refugeedom – leads to the presence of students from many different cultural origins in the same classrooms. In some Canadian cities (e.g., Montreal, Toronto, Vancouver) it is not unheard of to have children from 20 or 30 countries (cultures) of origin mixed in the same school, grade level, and even class. The most recent PISA 2009 report shows that there are schools in Canada with more than 60 percent of the students taking classes in a language other than their mother tongue. It is interesting, therefore, that in this age of increasing globalization, very little has been done to theorize the experience of the foreign/strange, which inherently lies beyond the (hermeneutic) horizon that encircles what any individual I (we) can understand. In this chapter, we ground ourselves in the work of the German philosopher Bernhard Waldenfels, who has worked for many years on the problematic that is at the heart of this chapter. Thinking about the foreign/strange is not only useful for understanding learning of a different culture (e.g., Korean versus Canadian or German) but also for thinking the inherent difference between the vernacular culture and the culture of scientific publications. In the German of Waldenfels, the central term is the noun "das Fremde," which denotes both the foreign and the strange. To reproduce the semantic field of "das Fremde" as much as possible, we use the concept of the "foreign/strange." We explicitly do not use the concept of the Other to denote the foreign/strange (other), as there are many forms of Other that are not foreign and strange. The Other in this chapter denotes the generalized other that is concretely realized as a second person – Emmanuel Levinas' "face" or Martin Buber's "Thou" – in the encounter. We frequently make use of the first person not to denote ourselves but the fact that in each case, an experience is "mine" from the perspective of and for a specific person. We thereby eschew a reduction of experience to the third-person perspective, which is the perspective of nobody or some suprahuman being.

In many instances, studies in science education presuppose that science is value-free and applicable independent of (cultural) context and therefore few have attended to the possibility that university students who have migrated from one country to another for the purpose of studying may have special experiences and needs to cope with their displacement. Yet, moving between countries and cultures may have effects on learning even the most abstract of the sciences and even in the most innocuous places. We exemplify such an effect by means of an episode that we noted in our ethnographic study of a physics course: a Japanese student (Mariko, pseudonym) attending a physics lecture in a Canadian university where she intends to prepare for a career in climatology. We begin with an episode from a thermodynamics lecture in which the professor (who also features in chapters 2 and 5) talks about some fundamental concepts of thermodynamics, including the Helmholtz concept of free energy. The professor has written some equations on the board (Figure 7.1) and talks about an isothermal process. SungWon, who also attends the class, subsequently writes into her research notes a commentary pertaining to this event.

Episode 7.1

Professor	((*standing toward students at the front of the classroom*)) in an isothermal process (3.9) and that maximum (0.9) is obtainable (0.2) only (1.0) in a in a reversible process ((*moving to the table and facing the chalkboard where he develops some thermodynamic concepts*)) how else can we interpret this Helmholtz free energy (2.3)
Mariko (student)	(then?) is it (be both?) isothermal and irreversible

Figure 7.1. A physic professor stands in the front side of the classroom. He faces the chalkboard where he develops some thermodynamic concepts.

SungWon: I remember one of physics classes I videotaped following Mariko, a Japanese student who majored in undergraduate physics. That day, Mariko was sitting, as usual, in the front row. She was attending to the lecture while making notes. I was at the back of the classroom, recording the class with a camcorder. The professor was talking about the equation of the Helmholtz free energy and the physical interpretation of it. The moment happened when he finished an utterance with a rhetorical question that sounded like he was orienting toward himself ("How else can we interpret this Helmholtz free energy?"). Mariko asked if the physical condition is both isothermal and irreversible – the two words are physics terms that are used to refer to the states of thermodynamic cycles. The question is one of those that any physics student might ask to confirm what she heard from the professor. But to me Mariko's question sounded very different from those that other students asked: It startled me and thereby became immediately salient in my mind. I felt that something important happened to her. I felt as if something happened to me, something struck me, and a sense of disquietude emerged from the depths of myself.

I knew that Mariko rarely asked questions in class. Once she found something to be unclear, she normally preferred talking to the lecturer after class or visiting his/her office individually. But here, she actually asked a question in class. To some other physics students, this behavior might look more appropriate: In the culture of physics, few students go to see the instructor because it is a sign that one does not cope. But from my experiences of having grown up in an Asian country (Korea) and studying abroad (Canada), I knew how difficult it would be to interrupt an ongoing lecture by one's question and to talk about something in the foreign language (English), a tongue that is not one's own. I remember experiences that I found myself having already lost a good timing to speak out once I figured out what I wanted to say. Therefore, when I saw and heard Mariko talking in this situation, I not only recognized that something uncommon had happened but also felt unsettled as if I had asked the question myself. (SungWon, field notes)

In this situation, Mariko is sitting in a university classroom and listening to a physics lecture with other (native, foreign) students. Migration in many instances is not easy, as it frequently leads students to be exposed to the foreign/strange (other) and to experience various forms of being-affected like the one that arises from speaking in a foreign tongue (see the beginning of a poem by an Indian-American-German poet in the sidebar). For example, the discomfort and disquietude Mariko felt about asking a question in class – because in her native culture (Japan) it is seen to be an interruption to the professor's talk and therefore an impolite form of behavior both to the professor and classmates – led her to be silent about that which was not clear to her; and, as she told us repeatedly during interviews, she thereby felt estranged from the class. The otherness (alterity) she experiences would delay attending to and dealing with the experience of the foreign/-strange itself. Here we define the foreign/strange as that which lies outside our horizons of understanding, outside that which we can grasp by drawing on the familiar and familiar experiences. At the very moment that we come to grasp something, its foreign/strange nature has disappeared – the foreign/strange itself has receded. From the perspective of learning science, an important aspect of relating to the foreign/strange lies in the withdrawal of the foreign/strange and its resistance to be captured by the familiar (discourse). Even though we learn, there remains more to be learned without that we can know what it is – and this is so even with the things we just learned or even though we might have taught the topic for 20 years (something we were told by two professors independently, who, even after having taught a particular subject matter for this period of time still feel that their understanding of it increases). Mariko's act of asking a question constitutes a significant moment of learning in this sense. Thus, we ask from the perspective of the learner: How do I learn in and through the experience of that which is unknown to me and therefore is part of the realm of the foreign/strange? "What would you do if you had two tongues in your mouth," writes the poet Sujata Bhatt from her own experience, "and lost the first one, the mother tongue, and could not really know the other, the foreign tongue?" How can I take hold of something that is not just *other* but that is so radically other that I cannot

> *Search for my tongue*
>
> Days my tongue slips away.
> I can't hold on to my tongue.
> It's slippery like the lizard's tail
> I try to grasp
> but the lizard darts away.
>
> (mari jeebh sarki jai chay)
> I can't speek. I speak nothing.
> Nothing.
>
> (kai nahi, hoo nathi boli shakti)
> I search for my tongue.
> . . .
> You ask me what I mean
> by saying I have lost my tongue.
> I ask you, what would you do
> if you had two tongues in your mouth,
> and lost the first one, the mother tongue,
> and could not really know the other,
> the foreign tongue.
> . . .
>
> (Bhatt, 1988, p. 63)

conceptualize it and therefore cannot intend? How do I learn when not only the subject matter content is foreign/strange and beyond my horizon but also when the cultural context itself is foreign/strange? The cultural context here matters because the clarification of something a physics (as any other science) professor produces for what he is saying usually is achieved by means of discursive devices that already are a part of ordinary language, itself deeply embedded in and constitutive of culture.

To explain and explicate the relation to the foreign/strange in general and Mariko's action in particular, most scholars use concepts such as intentionality and agency: A student appropriates the foreign/strange by learning it. But this approach glosses the difficulty of understanding *how* we can learn the foreign/strange if it lies beyond our (hermeneutic) horizon, that is, if it lies beyond that which we can envision, anticipate, and integrate with the already known. Scholars who provide the simple and perhaps simplistic answer explain that Mariko asked a question because she had a (strong) intention to learn. Yet, once we consider that there is an asymmetry in the first-person relation to the foreign/strange, such answers do not suffice and do not capture the phenomenon. Precisely because the foreign/strange denotes something that is beyond my horizon and therefore inherently unknown to me, it cannot be intended and therefore I cannot grasp it from my first-person perspective. How can I intend learning something that I do not have the means to select as object of my intention?

Beginning with Franz Brentano (1874/1924), phenomenological philosophers have made thematic the object-oriented nature of intentionality. There cannot be an intention without an object. Precisely because I cannot know the foreign/strange and because it lies on the other side of the horizon that includes what I can grasp, I confront this foreign/strange when I am exposed to the un-anticipated, unfathomable, and unknown. Heidegger (1927/1977) uses the idea of the clearing, within which there is light so that we can see but beyond which there is the dark that we do not know what it consists of and what it contains. This dark is the foreign/strange. Because I cannot anticipate the nature of the foreign/strange, I have no other option than opening up to be touched and affected by whatever it is. Only once the foreign/strange has affected me, can I begin to integrate it to what I already know. But as I appropriate something, it no longer is foreign/strange. That is, in the very instant that I appropriate something new, I have not appropriated the foreign/strange, because I have already approached it in terms of what I know rather than something in itself. The foreign/strange has not been appropriated but has receded and withdrawn thereby remaining inaccessible to my efforts to master it. I only can be exposed to the foreign/strange by opening my body to the unknown world, allowing my senses to do their work. My living body – living because it is characterized by intentions and affections – thereby is open to be affected by something that is *radically other*.

The German (and French) phenomenological literatures make the distinction between *Körper* (Fr. *corps*) and *Leib* (Fr. *chair*), both of which are *body* in English. However, the former term (*Körper*) refers to the material aspect of the body, its nature as material object in the way a coroner would find it, whereas the second term refers to the experiential dimensions of the body, which includes the senses. We therefore use the term *living body* to capture the distinction with the merely material aspects of the body. We use the term *radically other*, because that which it denotes is not merely another entity different from me but unintelligible to or before me: This other is radically other beyond all comprehension. But whatever I sense and whatever it is that I thereby make sense of – which always is in terms of the familiar – no longer can be the foreign/strange, because the radically other recedes as I grab hold of something in terms of what I already know. This encounter with the foreign/strange and the associated receding that occurs as I grab hold of something, point us to a temporal diachrony within one and the same experience. This experience comes to be shifted with respect to itself: being-affected (pathos) by the foreign/strange and responding (e.g., understanding) to that which no longer is foreign/strange simultaneously are two aspects of the *same* coin that also stand in a diachronic relation. These two moments, pathos and response, therefore constitute a dialectical relation that cannot be reduced to one or the other of its manifestations.

Learning something that is beyond my horizon of understanding is possible because, as other human beings, I am endowed with passibility (Roth, 2011b). But opening up in this way also means I make myself vulnerable, for I cannot know how the foreign/strange will affect me. I may have the intention to achieve or learn something, but the experience of the foreign/strange exceeds any intention I may have. Mariko asked her question but could not know what would return, for if she had known she would not have needed to ask. To find out, she has to open up and listen, thereby allowing herself to be at the receiving end of the professor's locutions, which affect her. There therefore is an essential, inescapable, defining and thus *radical passivity* at the very core of any intent to learn something yet unknown and impossible to aim at. Although Mariko might intend to learn, she cannot intend *what* it is that she will have learned, which she comes to know by being-affected, and therefore, in total passivity. Learning something new inherently involves passivity: the "passivity of the initially vaguely awakened (insight) and the eventually increasing clarify of that which appears" (Husserl, 1939, p. 211). Even the reawakening of the originally experienced is characterized by "the passivity of a subsequent consciousness, of the flux of the fading consciousness of that which just-has-been" (p. 211). Thus, rather than constituting a purely agential phenomenon, learning is fundamentally characterized by the passivity of the first constitution of knowledge and its subsequent reconstitution. It is a confrontation with the unknown, and therefore, with the foreign/strange.

The experience of exploring something by means of touch can be used to illustrate that the intention to learn about some surface involves both agency and radical passivity. Because of this, Mariko's actions in the lecture cannot be fully captured by means of the notions of agency and intentionality. This, then, leads us to the core of our current problem: the globalization of science education pushes learners into a double encounter with the foreign/strange. First, immigrants and migrants find themselves in situations where the cultural practices are different from the ones they are familiar with (as shown in chapter 6, we think these cultures in terms of heterogeneities); and second, they find themselves in situations where they are asked to learn science and scientific concepts that are beyond anything that they can imagine even if they heard about it in their mother tongue. The former situation is important, because ultimately, lecturers allow their students to enter the unfamiliar physics discourse by means of a translation into the vernacular language and culture, which themselves constitute the foreign/strange for migrants.

A central problem of learning science arises from globalization: theorizing the experience of the encounter with the foreign/strange from the perspective of the experiencing person who is being affected. What am I to do, a student of science, when I cannot know what I am asked to learn and therefore how to get to know the epistemic object unknown and impossible to aim at? We approach this problem by theorizing learning and globalization in terms of a phenomenology of the foreign/strange (Waldenfels, 2006). We ask: How can we think the foreign/-strange on its own without destroying it in the very instant that we appropriate and reduce it to the familiar? How can we engage the foreign/strange without neutralizing or even denying it in the way we engage with it? In this chapter, we develop a theoretical framework beginning with the concept of *pathos*, the experience of being-affected, which has emerged from a series of studies concerning the foreign/strange. Pathos means that something is befalling us, and this in a way that the something neither can be founded in an a priori *what* (content of experience) nor rested in an a posteriori *what-for* (intent of experience). The term pathos has in its semantic field the notion of suffering, feeling, emotion, and passion (Roth, 2011b). Pathos thereby precisely describes my experiences in the encounter with the foreign/strange and the sense that arises from it: I am affected by something unknowable and unknown, and something that withdraws in the moment I reach to grasp it. The concept of pathos provides us with considerable theoretical purchasing power for explicating cross-cultural experiences that arise in migration and cross-cultural experiences that arise from learning something such as science, with its discourses that are radically other (i.e., scientific) than the ones I have learned and are useful in the everyday mundane world (i.e., pre-scientific, un-scientific).

> (jo vadla hoat toh)
> Since I have lost my tongue
> I can only imagine
> there is something crawling
> beneath the rocks, now burrowing down
> into the earth when I lift the rock.
> . . .
> There was a little girl
> . . .
> but I can't think of her in English.
> (Bhatt, 1988, p. 64, 65)

In this chapter, we unfold the theoretical gains obtainable from the phenomenological notion of pathos in three steps. First, pathos is grounded in the movement of the *living body*, for only the living body is the ground of agency, the possibility for emotion, and the senses that constitute sense. Without senses, I cannot be affected nor can I be oriented toward a specific aspect of the world that affects me (pathos). This orientation, the attention to the foreign/strange, constitutes the central aspect of the living body's *response* to the radically other. Second, the passivity of my living body – which has been affected by the other because it is open to the world – endows me with the capacity of *feeling with and for the other* in concrete transactions. (We prefer the term *transaction* to *interaction*, as the former is consistent with the irreducible nature of social situation.) This move has consequences in that such questions of who I am, my identity, is as much the result of a transaction as it is a resource for it. The "me" in this sentence, therefore, is both a condition and a result of the event (transaction). Precisely because I am capable to experience pathos I also have the capacity for em*pathy*; and it requires empathy to be able to experience pathos. This dialectical relation that links pathos and empathy derives from the non-self-identical nature of the Self, which at its very heart is other to itself. Precisely because "I is an other," to draw on Arthur Rimbaud's famous diction, my pathos is the pathos of the other (*em*pathos). Empathy therefore also constitutes the very condition for our doing research on the learning of others, not only when they have arrived from another society and culture but also when they are confronted with difference (a subculture) at the very heart of their own culture (itself a non-self-identical hybrid). Third, in empathy the Self is confronted with otherness in itself (conscience) – my Self is being-affected in and by my engagement with the foreign/strange. Conscience is the passive relation of the living body to itself and is the ground of solidarity.

In what follows, we articulate and explicate our phenomenology of the foreign/-strange in the course of three steps and thereby present a theory of learning and globalization on the ground of our auto/ethnographic data. We begin with the living body that lies at the heart of the experience of the foreign/strange and learning through it: The living body touches the radically other in its movement and is simultaneously touched (affected) by this other in so doing. We show that the living body is the place of intercorporeity (heterogeneity) in which the foreign/strange is translated into the proper all the while maintaining *difference* from the beginning – that is, without reducing the foreign/strange to the proper. In this sense, thematizing a phenomenology of the foreign/strange comes to denote a philosophical project that articulates principles of responsible, answerable, and ethical praxis of science education.

TOWARD A PHENOMENOLOGY OF THE FOREIGN/STRANGE

To theorize globalization and to use it as an analogy for learning, we require an approach that allows us to capture the encounter with the radically unknown, that is, we require a phenomenology of the foreign/strange articulated from a first-person perspective. Precisely because learning theories are formulated in terms of intentions, we need to theorize how I experience the unknown while migrating or while attempting to learn something absolutely invisible and foreign/strange to me. Saying that there is something to be learned that is currently unknown to me makes sense only from a third-person perspective, which objectifies both learner and the learned. From my, the learner's perspective, the foreign/strange that I am supposed to learn always lies beyond anything that I can grasp. Our approach begins with the concept of *pathos*, which only exists in and because of my living body, and the associated concept of *response*, which, too, is performed by the living body. On the ground of a dialectical theory of identity that is grounded in a philosophy (ontology) of difference, the Self *is* the Other – this formula precisely is captured in the book title *Oneself as Another* (Ricœur, 1992). This also implies that pathos *is* empathy, which has consequences for the very possibility to do research generally and to research the phenomenon of the foreign/strange particularly. Henry (1990) notes that "only the living – absolute subjectivities – enters into a community, which is that of life" (p. 162). Moreover, "the essence of absolute subjectivity in as much as it is the pure fact of immediately experiencing oneself identically is the essence of ipseity" (p. 163). Without empathy, a communally shared pathos, we would not be able to research the experience in and for others. In pathos and empathy, the Self confronts the Other at the heart of itself, and therefore the experience of the Other is coextensive with the experience of the Self (*conscience*). I could not understand the experience of the learner unless I am endowed with the capacity of empathy – which I am because of the pathic nature of my being. On the side of researcher, therefore, the phenomenon appears as the double passage of the Self and the Other that sublate one another – the research project constitutes a passage to confront the Self, at the same time of which emerges an understanding of a particular moment of learning.

The Living Body Responds in Pathos

The encounter with the foreign/strange occurs in and through pathos, that is, the experience of being-affected. To grab a better hold of the concept, consider this: I touch some surface to find out about its structure (consistency) because I do not know what the surface is like. I would not have to intend touching it if I already knew its consistency. But because I do not know what it is like, I have to open myself to be affected by the surface at the very moment that I realize the intent to touch (sample) it. Thus, whereas I can intend to learn about a surface by touching it, I cannot intend that which I am to sense and the sense that I develop thereof. I can only anticipate what I will sense if I already know what the surface feels like, which is precisely what I reach out to learn. As soon as I experience the surface "as rough" or "like sandpaper," the foreign/strange has receded and withdrawn. The surface characteristics are part of the known and familiar, and the unknown, unfamiliar foreign/strange has receded. My response to pathos has been the evolution and articulation of sense. But this response is inseparable from the pathos that allowed me to make the foreign/strange recede.

Pathos, as the term already indicates, involves emotional aspects of experience that I have in my relation to the world, which constitutes a form of knowing the world I inhabit. It is the condition for consciousness to emerge from the initial experiences whereby the living body feels itself and thereby is constituted as an agential power that we denote by "I can" (Henry, 2003). Without pathos, the world cannot appear in consciousness, because phenomenality itself depends on pathos. This is therefore an important concept for the theory of globalization, the situation in which the phenomenon is more salient than when considered in one's own culture (which for science educators is a nested culture involving everyday experience and science in the classroom). In this section, we take a phenomenological approach to a migrating student's (Mariko) experiences within and across (heterogeneous) cultures and show how pathos constitutes a dialectical whole with response, that is, it constitutes an irreducible pathos|response unit. Pathos and response always go together, as my exposure to the other – listening, being available to being affected – is the beginning of my response diastatically shifted with respect to itself.

From a cultural-historical perspective, my (the learner's) emotion and motivation are closely related to the expansion of my action possibilities that my participation is anticipated to bring about (e.g., Holzkamp, 1990). Mariko, who is participating in a physics lecture in the opening episode, came to Canada to pursue a university degree. She wants to become a climatologist and her anticipation of studying climatology constitutes a goal by means of which she realizes the cultural-historically evolved activity of "getting a university education." Because climatology is quite interdisciplinary in nature, she initially majored as an honors student in geography and then decided to study physics as her second major. She also participated in colloquia of other departments when these were related to climatology. All of these actions were associated with an anticipated positive (emotional) valence because they promised Mariko an expansion of her agential power (agency, action possibilities). However, her student life in the physics department was not easy and brought about forms and contents of suffering that she did not anticipate and, because it was part of the unknown and therefore foreign/strange, could not have anticipated. When she needed help for completing assignments, laboratory reports, or examinations, she preferred consulting with instructors rather than talking to her peers. This had been an acceptable way of dealing with learning problems that an active student would exhibit in her native Japan. Moreover, in Japan seeing an instructor provides others with evidence that a student respects his or her teacher. Seeing the instructor when she faced difficulties was not a problem in her first undergraduate major. But after she shifted to pursue the second discipline in which to major, physics, she began to have a sense that she was acting in a culturally inappropriate way: She experienced strange gazes of others in the department and thought they were treating her as if she had intentions other than assistance in learning. Even if others felt that they did not look differently at her, understanding what Mariko does and intends to do requires us to take into account her experience of strange gazes.

Confronted with faces that looked at her differently than before, being-affected by these gazes from beyond the comprehensible, Mariko did not know the origin of her suffering and how to deal with it. She began to avoid other people and tended to study on her own. It was painful for her to sit in class and look at/listen to a professor. She attempted to act as if nothing had happened and thereby to get herself out of the situation; but the situation did not improve. She began to feel emotionally drained. It affected her study: she no longer was an honors student. Mariko found herself in a situation that she could not have predicted, because of the foreign/strange (physics) culture within a foreign/strange culture. As agent, I cannot ever predict, anticipate, or intend the absolutely (radically) other. I cannot therefore get ready to confront in physics and in Canada what inherently is unknown to me. I only can meet (undergo) it in passivity, by opening up and allowing myself to be affected. Pathos thematizes suffering, emotion, and feeling that come with passibility – my capacity for being affected, for suffering, and for having sensations – and *radical passivity*. Here, radical passivity transcends any decision (intention) not to act, which in itself would be a form of action: radical passivity is part of my experience of having intentions that I have not intended so that I can only be a host (hostage) to them. Pathos therefore is a concept that allows us to capture the suffering, emotion, and feelings Mariko experiences when she exposes herself to the physics culture that not only was heretofore unknown to her but also she cannot understand because it is still foreign/strange. Pathos means that I am befallen (affected, done to) *by* something (*affect* is derived from the Latin *afficere*, to do [*facere*] to [*ad-*]). But I am befallen (affected, done to) without being able to ground the experience in a preceding *what* or in a succeeding *what-for*.

Pathos denotes that something is happening (done) to me, that I undergo something. It *just* happens. The content of pathos not only is the unwilling but also it cannot be willed. I can will to be affected as I touch some surface, but I cannot will the content of my touching – this I have to undergo. Mariko wanted to study physics. But she could not anticipate what would happen to her in the process, for otherwise she, concerned with expanding her power to act, would not have engaged in an experience that stifled and threatened her and, in fact, decreased her action possibilities. This experience of something happening to me cannot be grounded in some cause–effect model or in a model based on intentionality. Rather, the special aspect of pathos is that someone engages with the foreign/strange that withdraws in the instance that I grasp it. Pathos and grasp are two diastatically shifted moments of the same experience, an experience shifted with respect to itself. This experience therefore involves a response (*respons*ibility [Levinas, 1978]) or an answer (*answer*ability [Bakhtin, 1993]), which, while not contemporaneous with pathos, nevertheless stands in a dialectical relation with it. Thus, in the opening episode, when Mariko asks a question in class, she responds to the foreign/strange. As a living being she is oriented toward dealing with something unknown that touched her. This response began prior to actually asking the question because to be able to hear what the professor says, she has had to open herself to be affected. She has to be passive in the very moment of active listening so that she could confront the foreign/strange in and with the passivity of her living body – responding begins when she exposes herself to be *af*fected in and by the physics lecture.

Central to the pathos|response unit is the living body, the capacity of passivity that it offers (passibility) together with the agential means to have intentions and act in ways that realize them. I am affected and have experiences only because I am endowed with senses and therefore with passibility, literally, the ability to feel and suffer (Lat. *pass-*, past participle of *pati*, to suffer + *-ibilis* (-bility), capacity for). It is precisely the openness of my living body to the world, its capacity to be affected (pathos, passibility) that I can understand anything at all. For, "[t]he world is comprehensible, immediately endowed with sense, *because* the [living] body, has the capacity to be outside of itself because of its senses and its brain" (Bourdieu, 1997, p. 163, emphasis added). Pathos, grounded as it is in the senses, gives birth to sense: it, too, has an essentially and radically passive moment (dialectical aspect). Whereas I might (intentionally) pursue the making of sense, I have no control over sense but can only accept or reject it when it happens upon me (Roth, 2011b). Mariko frequently has been telling us that she is not merely misunderstanding others. She recognizes in the gazes of other (cultural) members of physics – particularly those gazes suitably described by the adjective "cold" – the kind of gazes that she would orient toward persons very different from herself. She understands those gazes, and this understanding requires her to respond to the other by opening up and allowing herself to be affected. Her response has begun before the act of asking a question occurs in the classroom.

The Living Body Empathizes in Pathos

In the preceding subsection, we articulate how pathos|response arises from the movement of the living body. Pathos arises in the encounter with the foreign/-strange because of the living body, but my pathos is not just mine because the living body, through the identification of the Self as Other, always already is *singular plural*, both mine and not mine (other). As Derrida notes, the individual living body is singular ("no one can die in my place" [1995, p. 60]) but this singularity is *intelligible* only within a plurality ("as soon as one enters the medium of language, one loses that very singularity" [p. 60]); it is through the plurality that we come to understand ourselves as singularities. Thus, Mariko's pathos is a concrete realization of passibilities and possibilities that the generalized other has as well (plurality) because these are dimensions of life itself; and because of this, her pathos already presupposes the possibility for empathy. More so, this pathos|empathy dialectic lies at the heart of our (the authors') capacity to conduct this research project that, once we get into the field, apparently turns out to be an auto/ethnographic project.

SungWon: This research project was far from making observation and having description of it; rather it led me to form a strong sense of connection with Mariko.

Michael: This sense of connection, grounded as it is in the pathos|empathy dialectic, is co-extensive with the transference|countertransference dialectic at the heart of ethnopsychiatry, which inherently involves the encounter between the analyst and the foreign/strange, the ungraspable that leads us to experience anxiety – thus the title of a well known textbook for social scientists, *From Anxiety to Method* (Devereux, 1967).

SungWon: When I came to experience Mariko's painful situation, I felt perplexed and anxious. Initially I was asking her why she felt in that way and whether she talked to people about this; perhaps I wished it would not be the case, not a serious problem. However, I soon realized that those suggestions would not make sense to her. If there was anything she did not attempt to do, it was because she could not do it. She was confronted with a harsh reality that she had to deal with at every moment of life. I knew this because of my experience of studying abroad, doing research in a foreign country (Canada).

Michael: What you say makes me think that we have not yet pursued the question how your own perplexity and anxiety produce effects not only in you but in Mariko as well. The resulting transference|countertransference dialectic thereby introduces "disturbances" that are impossible to eliminate from the data and therefore are an integral part of any account – acknowledged, as here, or, more frequently, as unacknowledged mediations of and in the results of research.

SungWon: My stay in Canada gave me opportunities to look at the world and myself "from a foreigner's perspective." Before I came, I may initially thought that I just would do research as I did in Korea, using the English language as a different code. However, it was not like that; nothing was easy. Normal tasks of research such as accessing the field, talking to people about the project, and getting at their understanding were nothing like "things as usual." For example, going to the classroom and asking students' consent to my recording turned out to be a big challenge for me. Frequently, I did not know where to begin and what to address. I deeply empathized with Mariko's feeling of frustration about studying in Canada and about herself. I knew the sense that things were not coming along although she was attempting to do well.

Michael: And precisely her understanding of your empathy opened up possibilities for the kind of investigator-participant relation that you established in the course of the project.

The living body is affected at the encounter of the foreign/strange, but this foreign/strange something withdraws from itself as the living body engages with it and as the senses contribute to producing/finding sense. Pathos and response therefore constitute an identical event grasped by two non-identical moments (dialectic identity of the non-identical). Yet, from my perspective as the experiencing person, the identity of pathos and response does not appear as such. The experience of the foreign/strange emerges from the temporality between pathos and response, which articulates itself as an experience of the Self that is temporally displaced – e.g., I understand what happened to me after the fact – and shifted with respect to itself – e.g. I am other than myself (i.e., discontinuity of identity). Given that the withdrawal of the foreign/strange is constituted by a process in which the Other becomes non-Other in the Self, I experience the discontinuity of and within me. Empathy matters in this situation because it constitutes a moment of praxis that allows the transcendence of the gap between pathos and response.

Empathy is a key aspect of the dialectic relation of Self and Other that lies at the core of all consciousness (Lat. *con-*, with, + *scire*, knowing), and thereby constitutes the response conjugated with pathos. The term empathy, equivalent to the German *Einfühlung*, was introduced in the 19th century based on the model of sympathy. Etymologically, empathy and sympathy denotes feeling what another feels: empathy (feeling from *en*, in + *pathos*) and sympathy (feeling from *sym*, in + pathos). In this chapter, we use the term empathy in the notion of "feeling with and for the other." Empathy constitutes a moment of praxis in which I do not feel otherness and which thereby leads me to be open to the unknown other that affects me. In empathy, I reproduce the Other's experience of being-affected without necessarily experiencing the identical content. It is affectability itself that I experience in being-affected. It therefore constitutes an aspect of continuity of identity at the very moment of the discontinuity a person experiences while confronting the foreign/strange. Empathy is a problem for a (radical) constructivist Self, because it does not even have a model to build pathos, let alone *em*pathy or *sym*pathy. However, on the ground of a dialectic theory of identity, empathy is the condition for pathos, which itself is a condition for empathy. Given that pathos and response are two moments of an identical event, empathy constitutes a form of response.

SungWon: One day Mariko began to talk about a course she had began to take; she talked about the professor and one classmate. Mariko felt comfortable about the professor and she had a friend on whom she could depend. Mariko let me know that she often did an assignment with Cathy; she suggested to me to come and meet her. Cathy was a Canadian student also majoring in physics. She had moved to physics from engineering and therefore, like Mariko, was unfamiliar with the culture of physics. Cathy liked this course more than any other courses in physics or mathematics. I asked their consent to record their meetings and joined them. Their problem-solving sessions, which occurred irregularly initially, came to be regular events. The two mainly worked in the students' lounge located in the physics building. Here, they found it easy to go upstairs to see the professor when needed. They assisted each other and improved their ways of collaborating. I was there with a video camera to record their work, but I was not just as an observer. Once they grappled with a difficult problem for quite a time, I felt like assisting them by doing whatever I could do if I thought it would be of any help to their study. Once I found they were improving, I felt really happy. Mariko was engaged in creating a space not only for herself but also for the other.

Michael: As in the other episodes you related to me and our readers, we cannot think your presence without the mediational effects that transference and countertransference produce, both enabled by the pathos|empathy dialectic that we are articulating here.

Empathy is grounded in the passivity that positions the living body with respect to other living bodies in pathos. We can pry something from the foreign/strange to become the non-Other while the foreign/strange recedes, because our sensuous (because living) body allows us to empathize with other things and persons and thereby achieve the Other at the heart of our Self. Particularly, the fact that the temporality of pathos|response does not come into view from other theoretical approaches makes the role of the living body central in conceptualizing the experience of the foreign/strange. Thus, being able to ask for help indicates that Mariko has begun to resolve an experienced problem. Our living bodies allow us to empathize with one another and thereby to respond to the foreign/strange with the Other. Empathy allows humans to engage with the foreign/strange in a responsible manner. It therefore does not come as surprise that Mariko empathetically engaged in this class.

The Living Body as the Place of Solidary Translation

In the previous subsection, we show that the living, sensuous body allows us to empathize with the other in pathos. Empathy is related to the passivity that the living body takes with respect to the Other; but passivity pertains to the relation to the Self as well because of the dialectical nature of identity. The relation of the Self to itself in empathy goes well with results of recent scientific research. Neuroscientists reveal that observing someone else's actions activates the same part of the brain responsible for emotion (emotional mirrors [Rizzolatti, Fogassi, & Gallese, 2006]). A person who is empathizing with another is feeling pathos that she would have in reference to the effect her action has. This opens up possibilities for me to take an Other's perspective of (onto) myself and thereby relate myself to me. Philosophers call this self-relation *conscience* (Ricœur, 1992), a term derived from the Latin for knowing (*scire*) and with/together (*con-*). My *conscience* always is a conscience grounded in and shared with the Other. I arrive at my Self only by way of the mediation of and through the Other. The pathos|conscience dialectic established on the one linking Self and Other constitutes a central principle of ethics, which in our experience of doing research, maintains the criticality of auto/ethnographic study.

SungWon: Mariko initially told me that she was not a good "sample" from the physics student population compared to those who exhibited excellent achievement without depending on others. She articulated her concern that she might not make any contribution to this research project. She was depressed because of her emotional pain and suffering. From a research perspective, it appeared as a problematic situation. I had no idea how professors and students would feel once I told them I would bring a video camera to their classes. More so, I had no bearing on knowing how this research would affect Mariko – philosophers call this impossibility to know the results of research *radical uncertainty* (e.g., Moros, 2005). I wanted to do research on physics learning, but if the research would be of no help to her it would be better for me to do nothing. I wondered what I should do in this situation. I was in a double bind. Because I empathized with her suffering, I could not enforce anything that she might feel uncomfortable about, but because of the very fact that I empathized with her, I could not be blind to it – if she were not in the situation that requires someone else's help, she would not have wanted to participate in this research.

Michael: Again, I can feel the anxiety that the project has brought about in you. The ethnopsychiatrist Devereux (1967) writes that access to the essence of the observational situation is not the study of the participant but that of the observer. Any data of behavior science therefore should be threefold, the behavior of the participant, the disturbances produced by the observational activities of the observer – your introduction and use of video camera and interviews – and the behavior of the observer.

SungWon: The answer to the problem did not depend on me but on reality, what I could do with her in terms of research. Mariko and I were in such a situation that my research can be productive insofar as she overcomes her problematic situation and simultaneously that the improvement of her physic study depends on how I proceed with this research project. However, I could not anticipate how to proceed and therefore I hesitated. Mariko helped me in this situation. . .

Michael: . . . which is a way in which your presence has "caused" a disturbance that the participant Mariko picks up to come to your help, thereby enacting form of empathy and response.

SungWon: She talked to me about her suffering, but did not want me to develop antipathy (from Gr. *anti-*, against + *pathos*, *pathé-*, feeling) against the professors and students about whom she felt uncomfortable. She told me what class she would prefer to make available to me. She selected some laboratory classes and suggested to me joining her in attending the department colloquia. I might have thought that I could help her, but in fact she was guiding me to do research and ultimately to take another look at my suffering. As I attempted to understand the world from her perspective, this made me better understand myself; I became aware of the cultural nature of my suffering and learned respecting myself as the way I wanted to do to Mariko.

The foreign/strange (other) exists within the person: My living body (*flesh*), seat of agency and sensibility, differs from my (material) body, seat of passivity and origin of resistance. My experience of the foreign/strange constitutes a transformation by means of which I come to appropriate something from beyond my horizon, from beyond anything that I can imagine, and appropriate it to that with which I am already familiar. I am exposed to and affected by something that is foreign/strange – inherently inaccessible to me and beyond all horizons of the intelligible (Lat. *intelligēre*, to see into, perceive) and therefore beyond the anticipatable, which does not depend on what we know or want, that is, is independent of our consciousness. This highlights the role of the sensuous body in my experience and understanding. The living body – endowed as it is with senses, passibility, and the capacity to be outside of itself to be affected by the world – constitutes itself as the place of this translation. It is Mariko's living body that is in the encounter with the (social, material) other and experiences the gazes: these gazes are not just figments of her imagination but experiences arising from the fact that something other impinges on her. It is in, with, and through her living body that Mariko attends the physics course and exposes herself to the lecture.

The translation of the foreign/strange into the known/proper (other) occurs by the mediation of the other – conscience is the ethical self-relation I experience, which is mediated by the Other in this process. This translation makes central the hybrid aspect of identity and intercorporeity at an ontological level. The person to which something happens literally is a *sub*ject (Lat. *subjectum*, *sub-*, under and *jacere*, to throw, cast), being-subjected to the actions of the other (world, person). The foreign/strange presupposes the realm of individuality: In my encounter with the foreign/strange I differentiate myself on one side of the horizonal field whereas on the other side there is the *whereof* the differentiation. Yet, this realm of individuality and being-Self of the Self is distinguished from the same that is conceptualized from a third-person perspective. Once we conceptualize the foreign/-strange in terms of its continual withdrawal and resistance to the attempts to be appropriated to (in*corporated* in) the familiar, we note that for the experiencing person, this is an experience in which the Self becomes the non-Other (that which is not articulated by any known other and comes to be constitutive of the Self). That is, Mariko's act of asking a question in class involves an aspect of otherness that now comes to explicitly appear as a constitutive part of her Self.

This otherness was already there in some way, since Mariko exposed herself to being-affected in and by the physics lecture despite the gazes of the others. This leads to an other-mediated relation that is at the heart of the Self. It is because I am other than myself that I can open a way to thinking about the foreign/strange other. If the Self is other than self – i.e., non-identical with itself – then it both incorporates, and is constituted by, this Other. The hybridity results from such bricolage and métissage of Self and Other (Roth, 2008). Therefore, in translating the foreign/strange, the Self meets the Other at the heart of itself, and a person comes to understand the Self in the course of understanding the Other. Because my pathos is the other's pathos – *em*pathos – empathizing with the other comes to be equivalent to engaging with the foreign/strange that I am experiencing. The pre-ontological condition that makes empathy and pathos possible – the existence of life itself – also constitutes the ethical ground from which solidarity arises. It constitutes the ethical ground of the living body, which in turn constitutes the place of translation in which solidarity is achieved. On the ontological level, the Self that is other than itself appears as the concept of intercorporeity (Merleau-Ponty, 1964). Because of intercorporeity of the living body, the proper (flesh, living body) and the other (flesh, living body) are tied to one another. Both share in and are expressions of the same life. This means that I am implicated in a web of social relations as a singular living body (node) among other singular living bodies (nodes), all of which are manifestations of life characterized by pathos. The living body is the locus of this translation, where agency and passibility, culture and nature pass into each other, as well as the locus where the Self meets the foreign/strange (other) and, in the encounter, changes. The living body is both proper and foreign/strange, self and other, and therefore the source of the intercultural (difference).

EXTENSION: PHENOMENOLOGY OF CONVERSATION

In the preceding section, we articulate a set of concepts suited for theorizing globalization and learning science in terms of the phenomenology of the foreign/strange. We theorize the experience of the foreign/strange within the structure of pathos|response and conceptualize the role of the living body in three aspects: first, the living body engages by means of a pathos|response dialectic; second, the living body *empathizes in pathos*; and third, the living body constitutes a locus of translation of the foreign/strange in solidarity with the other. In this section, we extend our analysis to exemplify and further explicate our theory. We do so by presenting an analysis of a concrete conversational situation. We revisit the case materials (in extended form) of the opening episode and articulate our answer to the questions we raised in the introductory part of the chapter: How did Mariko come to ask a question in the classroom? How can we rightly theorize the emer-

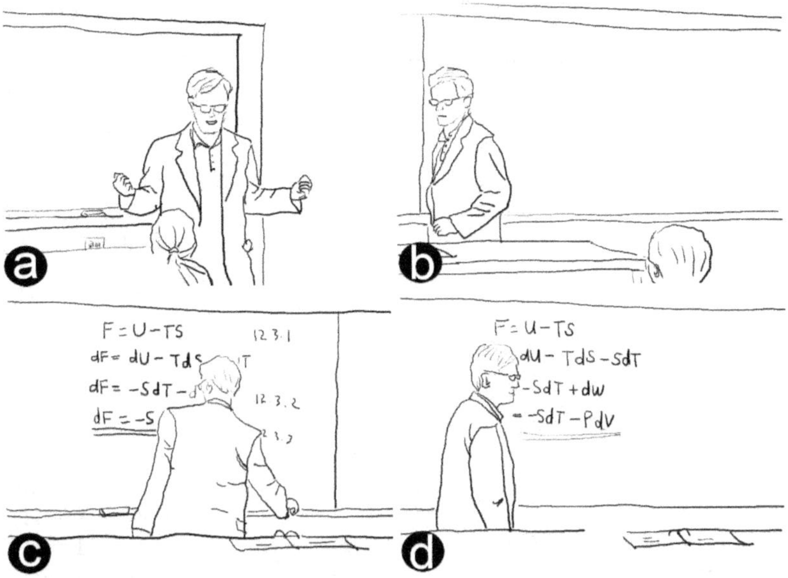

Figure 7.2. a. Cathy talks to the professor. b. Mariko looks at the professor moving to the table. c. The professor picks up a piece of chalk. d. He turns toward Mariko.

gence of her locution? We articulate theoretical perspectives that the phenomenological approach to the foreign/strange opens up for understanding the emergence of Mariko's speech in particular and structure of the conversation in general. Our analysis provides the material grounds for approaching pathos|response that students experience in concrete situations of learning science and associated emotional-volitional and ethico-moral aspects of conversation.

Episode 7.2 (opening episode expanded)

01 P: [the decrease in a Helmholtz free energy (2.9) [is the MAXIMUM work obtainable from system
[((*The professor walks from the right side of the classroom to the front with his face down. Around the right corner of the classroom he raises his hands up to the shoulders.*))
[((*The professor slowly turns toward the students and swiftly throws his hands down. Speech and gesture stop.*))
02 (2.4)
03 C: [<<p>in an:>
[((**Cathy raises her head up to the professor*)) [Figure 7.2a]
04 P: in an isothermal process↓ ((*The professor moves his gaze toward Cathy, who is having her face down and writing on her notes, and blinks his eyes.*)) and that maximum (0.9)

is obtainable (0.2) only (1.0) in a, in a reversible process
05 ((*The professor turns back and walks to the desk. *Mariko is looking at him. [Figure 7.2b] He looks at some part of his notes. He turns toward the blackboard and stretches his left hand to the left, where he cannot find pieces of chalk.*))
06 P: <<p> how else can we interpret this Helmholtz free energy>
((**He stretches her right hand to the right side of the blackboard and picks up a piece of chalk. [Figure 7.2c] He moves slightly left.*))
07 M: (then?)[is it (be both:?) (0.6) isothermal and irreversible
[((**The professor turns his body toward Mariko from the chalkboard.*)) [Figure 7.2d]

The professor walks at the right side of the classroom with his gaze oriented to the floor and utters "the decrease in a Helmholtz free energy." When he is near the right corner of the classroom, he stresses "is the MAXIMUM," while turning his body toward the students and moving his raised hands rapidly downward. He continues to say "work obtainable from system." He pauses. Cathy raises her head up toward the professor and, with low speech volume, says "in a." The utterance lasts longer than the usual, and thereby provides an opportunity for the professor to complete it. The professor immediately builds on the utterance and completes it: "in an isothermal process" (turn 04). He orients his gaze approximately in the direction of Cathy and winks with his eye. Cathy has already returned her gaze to the notes and has returned to writing something down. The professor continues: "and that maximum is obtainable in a reversible process." He then walks to the desk and looks at his lecture notes (turn 05). He turns to the chalkboard. Speaking with a low volume as if talking to himself, "how else can we interpret this Helmholtz free energy" (turn 06), he picks up a piece of chalk on his right side. His speech wanes and then pauses. After a while, Mariko raises a question, "Is it both?" and explicates, "isothermal and irreversible" (turn 07).

In this situation, the physics professor is talking about the definition of the Helmholtz free energy and the physical interpretation of it. He moves about the front of the classroom. He talks from the right side of the classroom and then walks up to the front. He stops and turns his body toward the students. Throughout those actions, he talks about the decrease in Helmholtz free energy as being equivalent to the maximum work obtainable from system. Not only the utterance but also his gesture, body orientation, and eye gaze are oriented to make the physics concept available to students. The action stops and brings about a pause in speech. Cathy who sits on the right side of the front row of seats, close to the professor, begins her utterance facing him. She says "in an" as if she continued the professor's preceding utterance. In fact, the proposition that the professor produced is incomplete because it makes sense only when the specific thermodynamic conditions are enunciated. The professor immediately produces what members can hear as a continuation of Cathy's utterance; and in this, he clarifies the condition in terms of "in an isothermal process." Moreover, while talking, he turns his gaze to face Cathy and gives her an eyewink, thereby making available his empathetic thinking and gives her an eyewink, thereby making available his empathetic thinking about the physical conditions. Participants can see in the eyewink an acknowledgment and a sign of complicity, expressing a shared sense not only of the content but also the mutual attunement to it. He goes on to say that "that maximum is obtainable only in a, in a reversible process."

In this conversation (turns 01–04), Cathy continues the professor's utterance and the professor, building on Cathy's turn, reifies and elaborates her utterance. Thereby, the transaction involving the professor and Cathy exemplifies a conversational instant that involves the Other as constitutive part of a person's action, and in the process allows empathy to be marked and shared: an exchange of empathy is materialized. From the perspective of the pathos|response structure, Cathy's uncompleted utterance "in an" (turn 03), which comes about after a long pause (turn 02) and lasts longer than usual at the end, indicates pathos, her being-affected by the professor's ongoing lecture talk in particular and something that she cannot grasp (physics) in general. Yet, at the same time, the professor's next utterance shows that Cathy's utterance constitutes the material ground from which his response comes about (turn 04). Her speaking is already a constitutive part of the response even before the professor explicitly speaks. As a result, the emerging lecture talk is co-produced, professor and students building on each other's contributions, which provide resources to articulate, elaborate, and explicate the concept of Helmholtz free energy. The professor's actions constitute part of response, rather than a source of pathos, to the otherness that Cathy and other students experience with respect to physics. Therefore, the two speakers in conversation are responding not merely to the other speaker but more importantly also to the Other, the generalized other, that the encounter makes available to speakers/listeners. The relation to the Other, an aspect of collective consciousness, is possible because the two speakers are materially attending to one another – i.e., Cathy articulates and the professor listens. The transaction that empathetically continues between speakers/-listeners constitutes a configuration of intersubjectivity that does not lead to estrangement but to the solidary translation mediated by the Other. The verb "to estrange" has the sense of "to cause to be strange, or a stranger, or as a stranger (to)" (OED, 2007). Estrangement and otherness are aspects of human life linked with and in part mediated by emotions, which therefore are concepts explanatory for the experience of the foreign/strange and learning as well. The conversation thereby produces an ethical effect that provides students with the possibility and opportunity for participating in publicly thinking aloud the topic of physics.

Pathos|response and the ethico-moral aspect of empathy constitute concepts for understanding the next situation (turns 05–07) as well. The professor goes to the desk, looks at his lecture notes, and turns toward the chalkboard. He gropes for a piece of chalk and speaks in a low voice, "how else can we interpret this Helmholtz free energy?" This utterance – which sounds like an act of thinking aloud – indicates that the professor clearly is affected, the pathos exhibited in his response to students. Yet, at the same time, his utterance as response is but the second, diachronically shifted moment of the event the first part of which is pathos – participation in responding to the otherness. The professor's talk brings about a pause. It

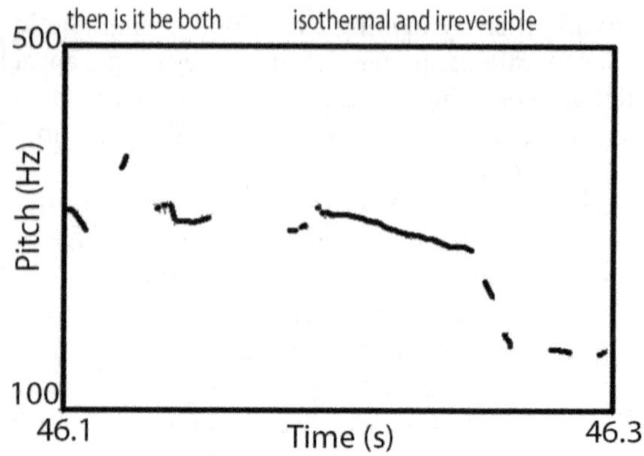

Figure 7.3. The prosody of Mariko's speech: the grammatical structure of the utterance shows that this is a questioning sentence, but the prosody shows different aspects. First, the prosody rapidly falls at the end whereas the pitch of a usual questioning sentence arises at the end. Second, before it began to falls (around the 45th second), the prosody is flat between 250–260Hz, which is higher than Mariko's average pitch (150–200Hz).

is precisely at this point, where the emerging pause provides an opening, that Mariko asks if the conditions are both isothermal and irreversible. Thereby, her act produces an effect of clarifying the issue and completes the professor's utterance as a request for students' participation in the ongoing clarification of the concept (Helmholtz free energy).

By attending to the lecture, Mariko and Cathy open themselves up to the experience of being-affected by something foreign/strange to them. The professor, whose role is to help students in learning thermodynamic concepts, also is in a situation of being affected while he is grappling with articulating an explication intelligible to students and to himself. In this situation, the professor's utterance and Mariko's subsequent question constitute a conversation that contributes to clarifying the issue of the ongoing lecture talk, and thereby realizes a moment of the collective responsibility for teaching/learning in this course. Empathy is central to the ethical effect of the conversation because in empathy the Self and the Other engage in a solidary translation of the foreign/strange. In speaking an empathetic utterance toward the professor, Mariko is in the situation of solidarity with the Other and the Self as well, which constitutes response to the foreign/strange (other). The initial high pitch level and the rapid descent at the end of her utterance show that the prosody of her utterance is different from a usual questioning talk (Figure 7.3); it makes salient the orientation toward the Other (as generalized other) and the Self. This episode thereby exemplifies how attending to the Other's pathos (empathy) lets me deal with my pathos because my pathos is the Other's pathos as well – like empathy, solidarity, the condition of being like someone else in some respect, here experiencing pathos|empathy, arises from the situation.

In the framework of conversation analysis, the minimum unit for understanding an action consists of a turn pair. Mariko's action constitutes an inseparable unit with the professor's previous action because the professor's action provides an opportunity for Mariko's action to come about – the two turns are in a non-deterministic possibility relation. In speech act theory, perlocution is the effect that an utterance (locution) that pursues a particular intent (illocution) has on the other conversationalist. That is, a speech act is distributed across two turns or a minimum of two participants, the one acting and the one being affected. The perlocutionary dimension of the speech act is exhibited in the turn that follows the utterance. The ethico-moral ground on which Mariko's action unfolds is its role as the perlocutionary (effect) component of the previous (professor's) locution "how else can we interpret this Helmholtz free energy?" This part of the speech act (locution/illocution) changes the ground that situates and allows the emergence of the remaining dimension of the speech act, that is, the effect the locution has on the listener (perlocution). An analyst may find this effect when the next action materializes in an utterance or some other action. Given that an analyst can never know beforehand what possibility will be enacted or even what constitutes the complete set of response possibilities, it is a discontinuous process. Yet, once we approach the issue from a phenomenological perspective, we find that this discontinuity inherently involves continuity.

In a conversation, a discursive action produces an effect, which becomes available in the subsequent utterance. But the second speaker speaks and thereby makes available an effect because she or he already has been subject to (undergone) the other speaker's action (locution/illocution). Action and effect are not temporally shifted, but are part of a diachronic experience that is shifted with respect to itself and distributed across conversationalists. The first speaker speaks intelligibly (empathy for the other) whereas the listener is open to be affected (pathos); when the second speaker responds, the situation is reversed (Roth, 2010b). The two-turn unit therefore implies a double pathos|empathy and pathos|response dialectic. Mariko's utterance (perlocution) emerged in and through empathetic transaction with the professor. The locutionary/illocutionary part of the professor's act is constitutive of Mariko's pathos, but at the same time, because locution always unfolds through transaction with the other, it is constitutive of response. Mariko's response has begun even before she speaks at this microstructure of pathos|response. Empathy constitutes a passage through which Mariko's speech act come about and is affected in doing so. Therefore, empathy is a passage through which perlocution (response) emerges, which at the same time constitutes a passage through which illocution (pathos) unfolds in an intertwined form. The living body is the place at which ethico-moral valence (perlocution) and emotional-volitional valence (locution/illocution) opens a passage to one another.

ETHICAL PRINCIPLES FOR SCIENCE EDUCATION IN AN ERA OF GLOBALIZATION

Globalization is changing the core issues of science education. The presence of students from culturally diverse backgrounds in one classroom calls the science educators' attention to cultural issues that would not have been salient otherwise:

the encounter with the foreign/strange. Migrating students encounter the foreign/-strange in science lessons in a double sense: (a) they are confronted with subject matter that they do not know and that uses forms of discourse substantially different from everyday ways of talking about natural phenomena; and (b) they are confronted with explications in an everyday language that is not their own and that embodies a different culture. It therefore does not astonish us to read that students who speak English as a second language often perform lower than their same-age peers, even if their subject matter competencies are high while speaking in their mother tongue. Science teachers, too, when teaching in culturally diverse classrooms, may find that their teaching practice does not stand on stable ground. Therefore, science educators need to attend to cultural issues; and they must not treat science language as independent of the everyday language in which ultimate explications are articulated (see chapter 6).

This chapter is a product of our phenomenological endeavor to "radically doubt" the current practice of science education (praxis) and associated ways in which the educational practice is talked about in theoretical terms. What is it like to come to know something that initially is completely foreign/strange, in fact, so foreign/strange that it cannot be anticipated? What deep-seated aspect of learning do students' experiences make available to us? Here we take a phenomenological approach to learning science and exemplify it with a case study of learning physics that a migrating university student experiences in a culturally unfamiliar context. Therefore, we conceptualize learning in terms of phenomenology of the foreign/-strange. We presented the pathos|response dialectic as a basic framework and articulated its diachronic and diastatic structure at two levels: narrative analysis at a macrolevel and conversation analysis at a microlevel. Our analyses show that the living, sensuous body is central to the experience of the foreign/strange: It is the seat of empathy and pathos, and, in the process, comes to respond to the foreign/-strange. Empathy is central throughout this translation of the foreign/strange to the proper because it opens possibilities for the Self to become non-Other without reducing the foreign/strange to the familiar and thereby respecting *difference*. The phenomenology of foreign/strange, which we approach in terms of the pathos|response dialectic, offers new possibilities to think of learning science: First, our theory pertains to what learners feel in their real lives (emotional-volitional value) and second, our theory deploys research praxis that is responsible for what learners feel (ethico-moral value).

Thinking globalization in science education in terms of the foreign/strange dialectic provides a framework for learning science not only as it happens to/is desired by migrating students but also for those who encounter the foreign/strange (the scientific) in their own language. Given that learning always means confronting the unknown, our theory can be extended to re-theorize science learning generally. In this sense, pathos|empathy, feeling with and for the other, is a mechanism that lies at the heart of learning. Nearly a century ago, cultural-historical psychologists suggested that questions of the mind (cognition, knowing, learning) could not be appropriately responded without considering the integral role emotions play in acting (Vygotsky, 1934/1986). Neuroscientists report that emotion is inseparable from cognition at the neuronal level. This chapter reveals a new dimension of emotions: They are central to knowing and learning because emotionality is the fundamental mode of *Being*, the way in which the Self is related to the Other – *pathos* constitutes the other side of *response* because pathos is empathy.

Empathy is a principle for the ethical praxis of education because it lies at the heart of collective responsibility and therefore the praxis that takes difference as the inner force of learning. From the perspective of a philosophy of difference, the particular always is a concrete realization of the general (e.g., science curriculum) and the cultural-historical development of the general is dialectically related to the extent to which particulars vary in concrete situations. As we show in this chapter, a particular action is a variation that involves the emotional-volitional relation to the other and is responsible for the effect that brings about on the whole network of particulars. Empathy is the fundamental condition on which particulars make differences in responsible ways with respect to one another and therefore come to produce a better proper, that is, the general that the responsibly related particulars concretely realize. In the concrete situation of learning science, empathy is a real power that allows various forms of communicative praxis to appear and therefore produce a heterogeneous set of scientific ideas. From a learner's perspective, the field of learning science is a place of solidarity: I come to understand that which is unknown to me when I empathize with the other and therefore when I am in a situation of solidarity with the other.

The foreign/strange actually leads us to our next topic: ethics and responsibility. This is so because an ethics grounded in the foreign/strange is incommensurable with an "ethical fundamentalism," such as Kant and the neo-Kantians have developed it; it also is incompatible "with a technological constructivism" (Waldenfels, 2006, p. 11). In addition, the response to a foreign/strange appeal begins prior to any answer, the contents of which could not be evaluated in terms of formal ethics. Responding begins with attending to, looking at, and listening to the Other. It is precisely this responsivity that stands for an answerability that precedes the actual responsibility for that which we deliver with the gift of the answer.

CHAPTER 8

THE BODY IN/OF RESEARCH ETHICS

> What is immanent to our sensible life and constitutes its being is truly the subjective being of movement which defines both the power of our body and the peculiar quality of our individuality. To be an individual is to have an absolutely original relationship with the world, and this not by way of an ethical decision at the end of a deliberately undertaken effort, but everywhere and always, in romantic exultation as well as in daily banality. (Henry, 1975, p. 106)
>
> The body exceeds the categories of a thing, but does not coincide with the role of the "body proper" that I dispose of in my voluntary act and by means of which "I can." (Levinas, 1971, p. 205)

In this quote from *Philosophy and Phenomenology of the Body*, the French philosopher Michel Henry makes a link between life, our subjective being, and the power of our bodies. This power, as Levinas points out in the second quote, also projects the body outside of itself as it affects others. The body therefore constitutes an original relation to the world, which is not one of conscious ethical decision, but an inherent condition – each one of my act effects the Other and other things. Sartre (1943) points out that ethics is tied to my very being, a body among bodies, that I constrain the Other and thereby find myself in an ethical relation. Thus, "from the moment that I exist I establish a factual limit to the freedom of an Other, I *am* this limit, and each of my projects traces the outline of this limit around the Other" (p. 449). It is not surprising, therefore, to find Emmanuel Levinas, one of the foremost 20th-century philosophers of ethics, to use a part of the living human body, the face, as the distinguishing moment of human existence as the synecdochical signifier of ethics in general.

Qualitative social research, designed to develop ways of understanding and explaining lived experience of human beings, is a reflexive human endeavor. It is reflexive: as researchers attempt to better understand their participants, they also come to better understand themselves. Consequently, research ethics itself becomes an ethical project, for it pertains to participant and researcher at the same time: Both are subjects, knower and known. Particularly in case of research on learning, reflexivity arises from the fact that the research itself constitutes learning about learning. How is ethics in research on learning reflexive of, in its praxis and praxeology, ongoing events and changes of the human learning? In this chapter, from our experience of conducting a project designed to inquire into "learning in unfamiliar environments," we develop pertinent ethical issues through a dialectical process. First, ethics is an ongoing historical event; second, ethics is based on the communicative praxis of sensuous bodies; and third, ethics involves the creation of new communicative configurations. We conclude that ethics is grounded in a fundamental answerability of human beings for their actions that arise in living labor, which requires communicative action that itself is a dialectical process in opening up possibilities for acting in an answerable manner.

ETHICS FOR RESEARCH ON LEARNING

> The ethical situation of responsibility is not comprehensible on the basis of ethics. . . . The tropes of ethical language are found to be adequate for certain structures of the description. (Levinas, 1978, pp. 191–192)

Fundamental to ethics in qualitative social research is the question about ways of understanding and explaining the lived experiences of human beings. Although it is quite recent to talk about research ethics at the institutional level, scholarly efforts in the social sciences – designed to develop various qualitative research methodologies that approach human knowing and learning on their own in the complexities of everyday contexts – have responded to the question in a wider sense. As salient in many social scientists' criticism of the artificial situations of the laboratory setting as a context for understanding human nature, objectivist approaches to human learning eliminate subjective self-experiences in the name of scientific objectivity and thereby produce knowledge incompatible with human subjectivity (e.g., Holzkamp, 1991): The living, sensuous human subject no longer exists either as the one we know about or the one who knows.

Common to some qualitative approaches is the fact that any pursuit of scientific objectivity necessarily involves researchers' subjectivities and the very act of constructing an object changes the subjective ground for action that has enabled that action. Thus, particularly in research on learning, some researchers have not resigned to describing and interpreting their phenomena but proactively dealt with issues emerging from the mutually constitutive researcher|researched relationship. They thereby value research not only for producing knowledge but also for contributing to human development and improving the human condition. That is, such qualitative research does not abstract from the lived experience of researcher and researched – which, despite claims to the contrary, even many qualitative studies do in creating observational categories that the participants no longer understand (e.g., Smith, 1999) – but inherently and continuously situates research and writing in the world we know through our experiences. Given that researchers' actions while inquiring into human learning changes the phenomenon – and thereby the researchers and their participants – two important aspects emerge concerning ethics in research on learning: First, the ethical value of research on learning is subject to a continuously developing research praxis that expands developmental possibilities of both the participating learner and the researcher in their interactions; and, second, the value of ethics as a description of praxis is subject to the development of ethical theories (deontologies) that are reflexive of human experiences and events coming about in the course of the research process. Concerning the relation between theory and lived experience, we prefer to say "praxis" to denote "its precedence to the theory that is used to describe and explain it" and also "praxeology" to denote "talk about or explanations of praxis, grounded in and developed out of praxis" (Roth, 2002, pp. 155–174).

How then is such an ethics possible, an ethics that is reflexive of – in its praxis and praxeology – the ongoing events and the development of the human learning under study? In the most general terms, the issue of establishing reflexive theories – those that develop in accordance with changing realities while providing explanations of that order of the change – has been a focus of social studies. Particularly in research on learning, there have been attempts to organize local events and theories through reflexive forms of research. In fact, these approaches have contributed to recognizing the importance of reflexivity in the developmental process of research. However, possibilities to achieve ethical theories in concrete research processes have been developed by phenomenological sociologists, who elaborated a reflexive approach to human lived experiences and the lifeworlds from third-person and first-person perspectives. "Lifeworld" denotes the functionally patterned world that an organism perceives and acts in within some activity; it is a real world objectively given to and experienced by a person – we do not normally doubt the world surrounding us. In conducting research, an investigator might take a participant's lifeworld as an object of inquiry and analyze it from a third-person perspective. But to understand how the participant experiences that world, the researcher has to take a first-person perspective of the participant's lived experiences as if they were her own.

Of significant importance in this approach is the reflexivity involving different forms of research praxis: first, praxis acting on another's experience in reflexive to one's subjective experience, and second, "phenomenological reduction" of subjective assumptions embedded in the praxis. In the case of research on learning, reflexivity arises from the fact that the research itself constitutes a process of learning about learning and may even lead to better understanding the subject matter (Roth, Tobin, Zimmermann, Bryant, & Davis, 2002). Yet much research on learning objectifies the human subjects it studies, the people whose knowing and learning is of interest; and it disregards the fact that researchers are the knowing subjects and, as learners, unacknowledged objects of study. As a result, we do no longer recognize the sensuous human being when we read about how a student solves problems in his science or mathematics class. With all the talk about cognition, knowing, and learning, the living being is expelled from research. Concerning ethics, reflexivity implies that the way researchers develop and conduct data collection and analysis bears a relation to the way research participants are developing as part of their learning activities under study. It is at this point that we see possibilities for addressing ethics as praxis and praxeology, an ethics that develops through human social interactions in concrete situations.

ETHICS AT ISSUE

Research ethics concerns the endeavor of configuring a human relationship in which a researcher and a participant can respect their different grounds for actions during their overlapping and interweaving collective activities. The exact nature of the relationship between a researcher and her participant is different from case to case. In the case of research on learning, however, particularly when the participant is a young child or student – who is often presupposed to have less knowledge and power than the researcher – the person's relationship to the researcher is quite different from other cases where the focus of inquiry is, for example, already demonstrated expertise (e.g., scientists, doctors, etc). In this section we articulate an ethical issue related to an asymmetry in knowing. The issue of who knows what during an interview can be very complex and therefore ought to be an empirical issue rather than be taken for granted. Thus, in a study concerning interviews an undergraduate student had conducted with expert scientists, it was shown that the matter of who was in the know was continuously contested (Roth & Middleton, 2006). At times the undergraduate student was recognized as knowing more about graphs and graphing than a professor in the physics department where the undergraduate student majored. As a way of developing the issue in sufficient complexity appropriate to the issue at hand, we first provide an episode depicting a critical moment from a study of learning physics and we introduce a reflexive narrative of the first author. We then develop the issues dialectically, that is, in a reflexive process.

A Moment in Research on Learning Physics

The following episode occurred during a regular three-hour optics laboratory class, one of the physics undergraduate courses at the Canadian university where the study took place. Mariko regularly attended the weekly class. The topic of her experiment on this day was "polarization" and included three subtasks: "The law of Malus," "Circular and elliptical polarization," and "Polarization due to reflection and Brewster's angle." The experiment was arranged in a small separate room apart from the usual bigger laboratory, because, in the case of polarization, it was important to keep the room dark or the sensitive instruments would not have produced accurate measurements. At the beginning of the experiment, the instructor came to the laboratory and gave instructions concerning the tasks and put various instruments on Mariko's laboratory bench. To conduct her first task (the law of Malus), Mariko was changing an angle between two linear polarizers (polarizer, analyzer) set in parallel between a light source and a sensor (Figure 8.1a) and measuring the intensity of a passing light by means of reading an intensity meter connected to the sensor (Figure 8.1b).

Episode 8.1

((*Mariko rotates a handle attached to a polarization plate with her right hand and gazes the sensor and the light intensity meter by turns.*))

01 Mariko: <<p>ninety degrees, this is fixed> *(18.6) <<p>yes, I think, [this (??)> (45.4) [Figure 8.2a]
[((*Mariko stands up, pulls a stool far from her, and exchanges it with her current one. After*

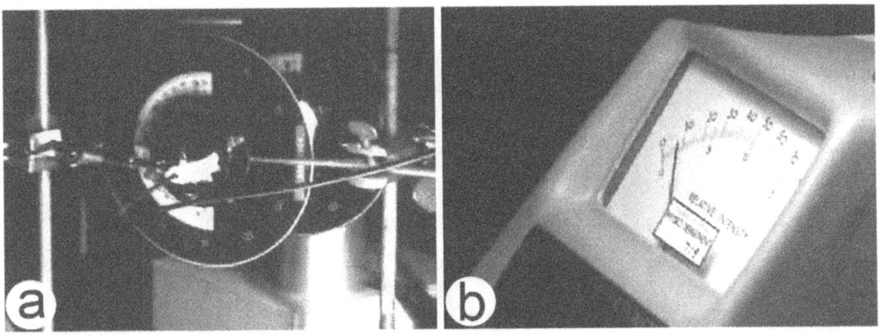

Figure 8.1. a. Two linear polarizers are set in parallel between an unpolarized light source behind them and a sensor detecting a light passing through in front of them. In the experiment, the latter polarizer is called a polarizer and the former is an analyzer. b. A light intensity meter is connected to the sensor. Seen on the scale is a fan-shaped graduation inscribed with two series of numbers, one above and one below. Two sockets (not seen in this picture) are found at the front bottom of the instrument.

sitting down on the new stool, she gazes at the polarizer and looks at the intensity meter in turns. While murmuring, she tries to write something down on her notebook but again gazes at the meter again.))
02 Mariko: Hm, (2.5), SungWon, [$_2$do you think *(1.5) either (2.0) either one is okay, I mean (2.3) [Figure 8.2b]
[((*Mariko points to the intensity meter with a pencil on her right hand.*))
03 SungWon: [Um

04 Mariko: ((*SungWon appears on the left side of Mariko*))
[*about this is ten, twenty, thirty, forty, but (1.4)
[oh- [Figure 8.2c]
[((*Mariko points to the numbers written on the scale with her left hand and SungWon looks at what Mariko point at.*))
[((*Mariko brings her left hand close to her face.*))
05 SungWon: the number
06 Mariko: number, [this should be=
[((*Mariko points her left hand to the numbers on the scale.*))
07 SungWon: [Here, there is a (0.6) range, you know, you can select (0.8)
[((*SungWon points to the numbers right above the two sockets with her right hand, denoting the range each socket can measures of written right upper two sockets.*))
08 Mariko: Um, (0.8) select (3.6)
09 SungWon: So [$_7$if you select this one,
[$_7$((*SungWon points to the left socket.*))
10 Mariko: um

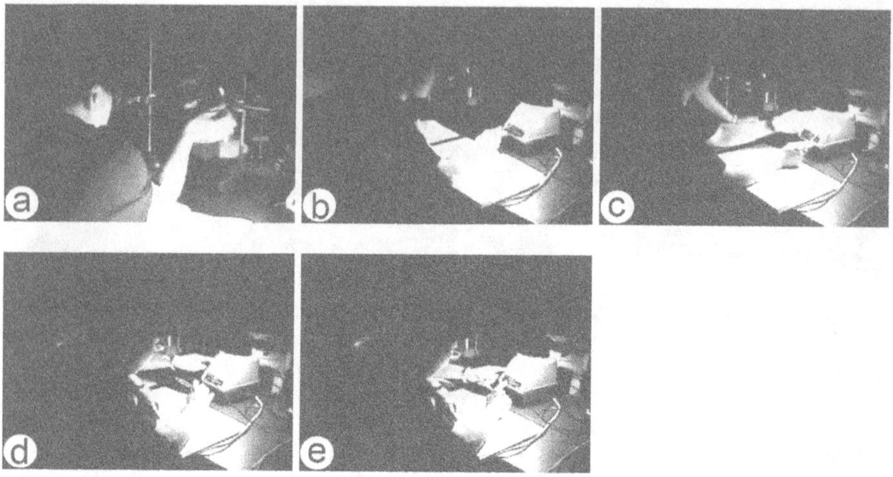

Figure 8.2. a. Mariko changes the angle of analyzer relative to the polarizer. b. Mariko asks a question while pointing to the light intensity meter with a pencil on her right hand. c. Mariko points to the numbers on the scale with her left forefinger. d. SungWon points to the numbers written on the upper part of the scale with her left forefinger. e. After mentioning the right socket, SungWon points at the numbers written on the lower part of the scale with her forefinger. Mariko points to the same end.

```
11  SungWon: then you read the number   [in the *upper: (0.35)
             [Figure 8.2d]
                                          [((SungWon moves her left
             forefinger around the numbers on the upper part of
             the scale.))
12  Mariko:  Oh, okay, okay
13  SungWon: [If you (0.57) select this one [then-
             [((SungWon points to the right socket.))
14  Mariko:                                 [*this one (??) [Fig-
             ure 8.2e]
15  SungWon: [you read the number here.
             [((SungWon moves her left forefinger around the num-
             bers on the lower part of the scale.))
16  Mariko:  Oh, that's great, thank you so much.
```

Description

In this situation, Mariko was carrying out the first task; she fixed the polarizing angle of the analyzer to 90° and thereby made its relative angle to the rear polarizer 0°. She gazed at the sensor and turned to read the light intensity meter (turn 01). She attempted to write something on her notebook, but turned her head up to the meter again. She uttered "Hm" and pointed her pencil in the right hand to the light intensity meter (turn 02). She called and spoke to the researcher, who was recording the experiment at the corner of the room, "SungWon, do you think, either, either one is okay" (turn 02). As SungWon came up to the left side of the laboratory bench, Mariko pointed to the double structured scale using her pencil (Figure 8.2b) and read aloud: "about this is ten, twenty, thirty, forty" (turn 04). She read until "forty" and then said "but." Instead of continuing, she kept silent for a while but for uttering an "oh," accompanied by a pointing gesture (turn 04). Her voice faded away, overpowered by the researcher's utterance "the number" (turn 05). Mariko repeated the word "number" and said "this should be" with her left forefinger pointing to the scale (turn 06). SungWon suggested, "Here, there is a range" and thereby cut off Mariko; she pointed to the labeled numbers written right above the two sockets, which denoted appropriate ranges of measurement (turn 07). As SungWon said "you can select" (turn 07), Mariko repeated "Um, select" and saw the researcher turning the two sockets (turn 08). After a pause, SungWon told Mariko "So if you select this one" (turn 09) and "you read the number in the upper" (turn 11), thereby relating the left socket to the numbers on the upper line of the scale. Mariko responded, "Okay, okay, ok" (turn 12). While pointing to the other socket, SungWon continued by saying "if you," which was followed by a pause and an utterance of "select this one" (turn 13). Mariko repeated: "this one" (turn 14). The researcher pointed to the lower line of numbers on the scale and said "you read the number here" (turn 15). Mariko responded "Oh, that's great, thank you so much" and thereby closed the conversation (turn 16). SungWon moved away, and Mariko looked at the meter by herself.

SungWon's Narrative

The laboratory instructor came to Mariko and explained what to do with the instruments and equipment. In the same room, I was recording the entire process with a camcorder, including their conversation. Once the instructor had moved away to take care of other students in another laboratory, Mariko began her first task. She looked carefully at laboratory manuals and instruments on the bench, particularly the different parts of a light intensity meter including the sockets on the front bottom and its scale. She turned to me and asked, "do you think either one is okay?" The instrument seemed to have confused her about how to select and what is a relevant socket in this task. She was asking me for help. I knew that selecting a relevant socket in this kind of instrument could be a confusing moment for a person who uses the instrument for the first time – I had similar experiences while studying physics and saw students have difficulties while teaching at the secondary school level. Nevertheless, at this moment I hesitated for a while, attending to a fleeting thought whether it would be appropriate for me to participate in her experiment by giving assistance.

I do not remember exactly what I was concerned about at that moment, but perhaps I was feeling a kind of conflict concerning what would be a relevant action as a researcher. Even before I had sufficient time to mull it over, I found myself moving toward Mariko's laboratory bench and explaining to her why there are two sockets on the meter and how each one corresponds to the two series of numbers on the scale. I returned to the camcorder and continued monitoring the recording. Almost at the end of the first task, Mariko appeared dissatisfied with the measurements she had recorded. She left the room searching for the laboratory instructor and returned with him. Soon, they figured out that she had not calibrated the light intensity meter and Mariko decided to repeat the first task with a calibrated meter. This situation reminded me of the previous situation where I felt torn: Would it have been better if I had helped her more carefully? At the same time I doubted: It would not have been the relevant action for me – researcher and physics expert – to assist a student in her task. Preoccupied with the research process, I began to attend to other than my ethical doubts. Almost at the end of the three-hour experiment, a worse situation occurred. I found that Mariko had not finished the three tasks within the given time and she felt bad about it. I was reminded again of the events at the beginning of the session. I began to feel responsible: I had refrained from helping her and now she had not finished the experiment within the allotted amount of time. What should I have done when she asked me for assistance?

Articulating the Issue

In the previous subsection, the immediate issue raised in SungWon's narrative is whether a researcher has a responsibility to assist a student research participant when the latter asked for help, and to what extent, if at all, she had an ethical obligation to assist. Sartre (1943) formulates the ethical issue in this way: "Charity, *laisser-faire*, tolerance – even an attitude of abstention – are each one a project of myself that engages me, and that engages the Other in his acquiescence" (p. 449). That is, SungWon's assistance may be seen as a charitable act, and abstaining, too, comes with an ethical dimension. As pointed out before, this kind of situation is not rare in qualitative research on learning where researchers already are knowledgeable practitioners regarding the topics and skills that the students in the study are supposed to learn. It goes without saying that being or becoming competent in the practice under research is important to develop a first-person understanding of the situation. However, without a reflexive step to overcome the prior understanding a researcher brings to a phenomenon, the first-person understanding is far from producing relevant knowledge. In this section, as a step toward a reflexive understanding of research ethics, we engage and elaborate a series of issues.

One ethical question emerging from the situation might be, "What do I (researcher) do when I am more competent at the practice than my research participant?" In a different context, the second author (Wolff-Michael Roth) and his collaborators have chosen a radical response – the process of research must be, in the first instance, of benefit to the teachers and students involved (e.g., Roth, Lawless, & Tobin, 2000). However, to respond to the question here we ask ourselves a reflexive question: "What presupposition does this questioning involve?" The presupposition might have been captured in one type of question: "Is it worthwhile for the researcher, as a person who collects data, to participate in a learning activity that she wanted to understand?" This question raises concerns about the quality of data. Or it might have concerned about the participant's learning: "Would the researcher's action be really helpful as she have intended? Would it not make the participant dependent on the researcher?" In any case, the questioning seems to presuppose that there is something incompatible between the researcher's activities of conducting research and the participant's activities of learning science to a certain extent and it looks so indeed. The critical moment must have arisen from the fact that Mariko's learning activity took place in the context of the research. Thus, in the optics experiment Mariko's every move is not only an act toward learning physics but also a contribution to the research; in the same way, the researcher's every move is not only an act of carrying out the research but also mediates the participant's learning. The two activities occur simultaneously in the same setting and unfold through interactions between researcher and participant.

It is important to note that this kind of situation occurred at the beginning of data collection but disappeared later. Although the researcher and the participant have never had a conversation about the event (i.e. which SungWon might try to have the next time in such a situation) or about the rule (i.e. how to act subsequent to similar situations), Mariko never again asked SungWon to help in the laboratory tasks, and SungWon did not to try to assist Mariko by getting involved. The critical instant occurred as both had learned at this moment how to act toward and respect the other. Therefore, we come to ask if there could be an answer to the initial question, or, in other words, if it was a relevant question. We propose that a relevant ethical question drawn from the episode should be about a moment situated in the historical change of praxis, that is, "How do the researcher and the participant experience contradictory situations and learn to act with respect to one another?" In what follows, we develop a response to this question by working through an analysis of the episode, which leads us to three assertions about research ethics. The historical change of collective praxis emerges not only over a long period of time, which is discerned retrospectively, but also in the short run, which is in the present tense. We can see it in various manifestations that human bodies produce in and for communication. We articulate the associated micro-processes in the fourth section.

DIALECTICS OF ETHICS

In this section, we analyze the interactions between the researcher and the participant at the microlevel and thereby elaborate our descriptions of the ethical aspects of research praxis and associated theoretical claims for ethics from a dialectical perspective.

Ethics as Ongoing Historical Event

In understanding the episode, one may say that the situation is simple and involves nothing important; the student asked a question concerning her instrumentation of the light intensity meter and in response to it SungWon went to the laboratory bench and told her how to read the scale. But, in her narrative SungWon the researcher described that she felt a conflict when she was asked a question, and responsibility when she came to know that Mariko's had used the meter without calibration and had to repeat the task. This subsequently resulted in not completing the task within the three-hour laboratory. Why did SungWon have to feel the conflict with and the responsibility for her action of attempting to assist her? This conflict itself is an indication that the issue was salient, because it has arisen in the research process, and, in its very possibility as a form of being, is also a possibility to other researchers. Here, the issue is not "what the action does in doing so" but its relation "to what it brings into effect," which unfolds temporally. Whereas the former is the present, the latter is the future; or, whereas the latter is the present, the former is the past. Produced in this relation is the fundamental uncertainty in ascribing the latter to the former and its agent, because a person always acts without knowing what is coming up in a future tense. Therefore, people, when they look back on their past actions, often make seemingly different statements. The researcher in the narrative might say later on "It should have been done differently" but at the same time argue that "It could not have been otherwise."

An action takes place in specific social and material settings, which provide a range of possibilities for action. In its unfolding, an action always concretely realizes one of potentially many possibilities; at the same time, all other possibilities remain unrealized. What was left out constitutes a background of action, that is, what was not salient "in doing so." However, an effect of an action and its ascription arises not only from what is enacted but also from the background because they both constitute "what an action does in doing so." Other possibilities might become salient after the subsequent situation has changed the ground of the ascribing act. An action a produced at time t_0 – i.e., a_{t0} – will be perceived differently after the fact: at some time t_1, it will be seen as $(a_{t0})_{t1}$ (Husserl, 1928/1980). Some time later, it will again look differently, as it is not a_{t0} that is present to me but a_{t0} seen through the way I have seen it at t_1. Therefore, the action will now appear as $((a_{t0})_{t1})_{t2}$ and will change each time I look back at it. As this formalism is cumbersome, it is better to write $a_{t0\text{–}t1}$, where it is understood that at t_1 the experience a_{t0} is apprehended through the thickness of the intervening window of time $t_1 - t_0$. Thus, descriptions that involve ethical evaluations such as "it would have been better if SungWon had checked the instrument calibration or if she had advised Mariko to ask the lab instructor" are possible only after we know that Mariko had to repeat her task and therefore could not finish it in time. The possibilities were not salient beforehand because SungWon now has a different interpretive horizon, one that historically evolved after the events. SungWon could not "recognize" the ethical dimension because time flows from past to present, not from present to past, and future episodes are emergent rather than determined by a determinate present. This, from our perspective, is the fundamental problem in much of public life in the attribution of responsibilities for events that were not and could not be foreseen. "Monday morning football" and Whig history are not good models for doing science.

SungWon's action of attempting to assist Mariko does not determine the forthcoming action nor is it completely unrelated to it. The relation between consecutive moments is always incomplete and changeable. There seems to be a gap between SungWon's concern for Mariko and the forthcoming situation, that is, between what an action does in doing so and what is ascribed to it as having brought about. A note of caution is appropriate here. The gap becomes an issue only when we consider it in the ideal world of possibilities. Despite the uncertainty, researcher and participant interacted and communicated with one another and thereby continuously reconstituted their grounds for actions. Two implications emerge from this. First, researcher and participant interact by means of resources available "here and now" and salient to themselves. Second, both can see new possibilities only when they do something and thereby come to stand on a new ground for action, which has been changed by the previous action. Thus, the gap does not exist because they respect their lived experiences. They actually know how things had to occur in that way, before talking about whether what they had actually done was right or wrong.

Therefore, fundamental to our first assertion for a praxeology of ethics is the historical structure of actions:

> *Assertion 1 Ethics does not exist as something independent of the lived-in world but only in the form of relations under specific circumstances of human activities. It arises from the temporal relation of (a) what an action does and (b) what is ascribed to it as bringing into effect. However, the action itself changes the ground of ascription and thereby the description of the action (discontinuity between the possibility for acting in some way and the factuality after having acted). This makes for a continuous process of development, which is open to historically new forms of praxis and subject relations.*

Research ethics arises in the endeavor of developing human ways of inquiring into human beings – researchers who reduce human beings to blips in correlation graphs, outliers, and aggregates do not face ethical issues in this way. Ethics is an aspect of social praxis, enacted by a researcher as a subject, which affects and changes the lived-in world rather than being a move in a world of discourse. Even the situation where a researcher or participant raises and considers ethical issues occurs in a specific social and material situation. Our goal is to do research that recognizes that the social comes into being in and through the doing of actual people (researcher, participant) under definite conditions not all aspects of which are under our control. This is a stance that opposes postmodern discourse, which has abstracted knowledge from the lives of people, and in which the subject (researcher and participant) exist only as characters in plots (Smith, 1999). What therefore really matters in ethics is to understand "how each interaction between a researcher and a participant constitutes a configuration of humanness." That is, the issues arising from our first narrative cannot be understood without also analyzing the concrete interactions between researcher (SungWon) and participant (Mariko). Of particular importance in this approach is the role of our bodies, because these are the very tools by means of which we articulate and make ourselves available to and for others. One of the ways in which we communicate is by means of the prosodic features of the voice, which modulate and mediate the words that others extract from the sounds we produce. In the following, we use the analysis of prosodic features to elaborate what we can learn from the transcribed episode.

Ethics as Communicative Praxis of Sensuous Bodies

In the episode, Mariko was setting up instruments to conduct her first task while muttering some words in a low voice (turn 01). She fixed an angle between the polarizer and the analyzer at zero degree and gazed at the light intensity meter. Her right hand was ready for recording a measurement. She looked at her laboratory notes where she had drawn a table to be filled with a series of intensity measurements and repeatedly attempted to write something at the first line of the table. However, instead of recording a measurement, she said "Hm" and starred at the light meter again (turn 02). The steeply falling pitch contour in her voice (Figure 8.3a) and her hesitating hand movements manifested uncertainty and insecurity. With and through her body, she was making her experience available to SungWon. Soon, her right hand pointed the pencil toward the light intensity meter that she was gazing at, thereby indicated that the issue was in the meter (turn 02). She called SungWon and continued speaking, "SungWon, do you think" (turn 02). Here, Mariko produced an utterance – an inherently *corporeal* act – in which the pitch contour had a structure of asking the other's response and therefore an action having more communicative feature than her previous actions (Figure 8.3a). There is no other way to communicate than with and through the body. We specially mark the *bodily* nature of the production, for even most qualitative research writes the body out of its accounts and features talk as disembodied discourse only (Smith, 1990). It did so actually. The researcher responded to Mariko's calling with a short "Um" (turn 03) and began to move from the place where she set up her camcorder directed toward Mariko's laboratory bench. Mariko moved her pencil right and left along the bottom of the scale without saying anything (Figures 8.1b, 8.2b). After the 1.5-second pause, Mariko said "either" (turn 02) and thereby explicated that the issue was related to some alternatives in the light meter, but stopped talking again. She moved her pencil to the right side of the scale during two seconds and then said "either one is ok" (turn 02). Here, she seemed to complete her questioning; but immediately she made an additional remark, "I mean," with a long flat pitch line (Figure 8.3b). The action produced an opportunity for something to follow and what followed actually was a pause of 2.3 seconds. In the meanwhile, the researcher almost had arrived at the laboratory bench. As soon as SungWon appeared on Mariko's left side, the latter continued talking: "about this is ten, twenty, thirty, forty" (turn 04). Mariko read aloud the numbers while pointing to them using her left hand. Now she articulated with and through her body what to see on the scale in relation to her issue in more elaborated speech and gesture. Furthermore, she produced another opportunity to articulate it more, rather than completing it there, by saying "but" with a high pitch compared to a declining pitch contour preceding it (Figure 8.3c).

In this situation, Mariko was articulating what she perceived as problematic by elaborating structures in speech and gesture in and through her body. Mariko opened by saying "do you think," but the object of "think" was not clear from the beginning. Rather, it developed as her communicative actions unfolded, from "either" to "either one is okay, I mean," and again "about this is ten twenty thirty

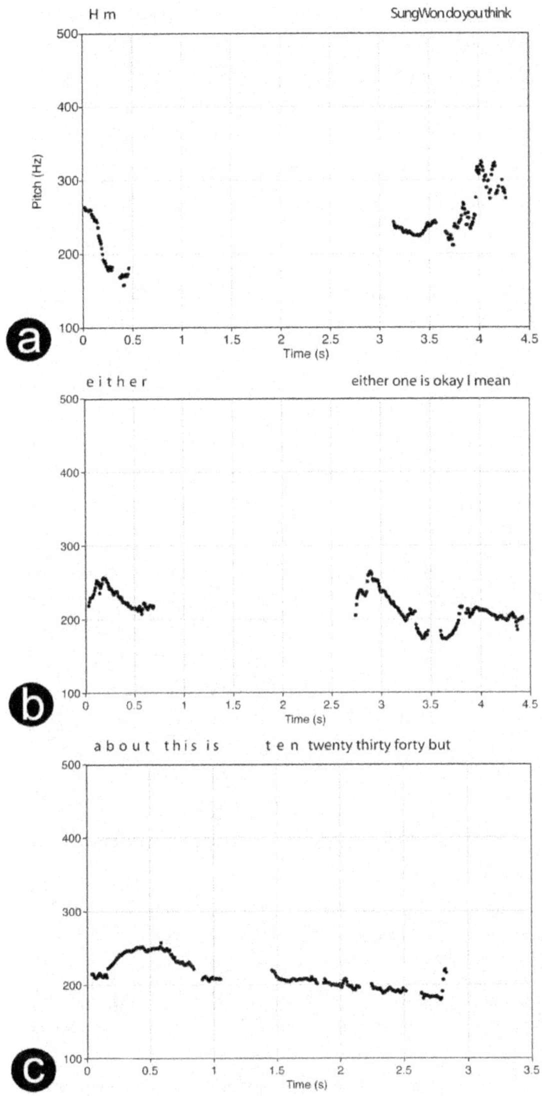

Figure 8.3. a. In our analysis, Mariko was found to make the "hm" sound frequently in different pitch contours. In this talk, "Hm" is close to a sigh of unpleasantness. The second part, "do you think," has a pitch contour that makes it as a questioning sentence requiring an object. b. The pitch of "either" extends a little while at its end and thereby lets the following long pause unsurprising. The second part, "either one is ok" has a pitch structure of a complete questioning, but the subsequent talk, "I mean," at a same pitch for about 0.5 second makes the sentence incomplete again. c. The gradually decreasing pitch rises abruptly at the end, thereby makes the whole utterance uncertain.

forty but," involving pauses between them. At the same time, Mariko's gestures in and between locutions had changed. At first, she just pointed the light intensity meter (Figure 8.2b), but she made a next movement of the pencil along the bottom of the scale. Then she pointed the pencil to the right side of the scale and finally even came to use the other hand to point a series of numbers (Figure 8.2c). Whereas she had began producing an utterance that could be heard as a question, she actually co-articulated the situation by and for herself in a communicative form; she did not interrogate SungWon. The elaboration became available to the latter (and her camcorder) because Mariko made communication materially available in the situation. In the three utterances and between pauses, we see their ending pitches constitute contrasting structures to preceding pitch contour in forms of a low (first), flat (second), and high (third) shape (Figure 8.3). Incomplete endings existed not only in the grammatical structure (what was said) but also in the physical structure (how it was said), which thereby produced uncertainty.

The communication provided flexibility not only to the speaker herself (Mariko) but also to SungWon who was coping with her question. Mariko physically produced time for the researcher to go to the laboratory bench to see what Mariko referred in terms of "either" and what she was pointing, and again the researcher's response allowed time for Mariko to articulate her issue. Although the video does not show how SungWon moved to the laboratory bench and therefore what kind of interactions occurred between the two in the meantime, there is some evidence that Mariko had attended to the researcher's movement. Only after a long pause Mariko uttered the words "about this is"; it was immediately after SungWon had come to her left side. Mariko did not look at SungWon directly but was aware of the other's moving toward her. Mariko acted in consideration of the time required for SungWon to arrive next to her. SungWon did not talk while she was moving but took a relevant position and timing in her communication with Mariko. Throughout these different modes of communication, the two were articulating the unfolding situation together. All the manifestations that their bodies produced (e.g., high pitch talk, pause, and gesture) constituted resources for evolving relevant responses to one another. In this sense, we see Mariko's initial actions constituted more like a request to the researcher of participating in her thinking and the two persons' communication constituted a process of collective elaboration rather than a mechanical process where one asked a question and the other was supposed to give an answer to it.

Therefore, the role of human bodies in communication is the central and fundamental element in our second assertion for a praxeology of ethics.

> *Assertion 2 Central to ethics is the role of human bodies that assume the (illocutionary) act and at the same time produce the (perlocutionary) act, thereby making an ethical commitment to and having ethical responsibility for the action. Ethics is therefore distributed over acting human bodies in the communication that constitutes the phenomenon both cognitively and emotionally.*

Ethics matters for human bodies and actions they execute are interpreted in terms of intentions; attributing an action to an individual (an agent) differentiates Self from Other. Central in this phenomenon of perlocution is the circulation of the act that human bodies produce and assume in interaction. Locution, illocution, and perlocution are terms used in speech act theory, each of which denotes "performance of an act *of* saying something" (locution), "performance of an act *in* saying something," and "performance of an act producing some consequential effects" (Austin, 1962, p. 99, original emphases). In this chapter, we use the framework not only for theorizing speech acts but also for any human act when it comes to ethics because it is of no use to say ethics without human communication or the communicative aspect of human action. The communicative action constitutes one's corporeal response to another's actions salient in and to the sensuous body. In the process of responding, the sensuous body acts on an illocutionary aspect of the action and assumes it, whereby the body produces a new aspect of the action that goes beyond its illocutionary aspect (perlocution). Here, what is cognitively and emotionally achieved with respect to what had been intended implies an ethical commitment to or responsibility for the action. The acting human bodies manifest commitment in the process of production and acceptation through their structured features (e.g., speech, gesture), thereby constituting a form of humanness, such as concern for one another. Ethics, in this sense, is distributed over human bodies involved in coproducing communication.

Ethics in Creating New Configurations of Communication

In the first part of the episode, Mariko articulated her issue in more elaborate forms in the process of speaking to the researcher who in the meantime could have come to the laboratory bench in response to the action (turn 01–04). In the next situation, the researcher was looking at the light meter while standing on the left side of Mariko. After her utterance "about this is ten, twenty, thirty, forty, but" and pointing gestures, Mariko was quiet. Then, all of sudden, she said "oh" as if she had come to see something she had not done before. Mariko's subsequent utterances quickly came about, but the researcher's louder voice "the number" overlapped it almost at the same time (turn 05). From a communicative perspective, SungWon's utterance suggests that the current issue was related to the numbers on the scale by rephrasing Mariko's speech and gestures in terms of "the number." The researcher's action was assuming what Mariko may want to indicate *in* her action (illocution) and thereby producing an effect of collective understanding. In as far as Mariko wanted to speak to the researcher about her concern, researcher's response explicated what had been achieved. In response to the utterance, Mariko repeated "number" and continued to say "this should be," while pointing her left hand to a specific part of the scale (turn 06). In so doing, Mariko expresses an assumption directly implied in the researcher's previous utterance and makes it a ground for her next action.

Intended or not, the researcher's utterance "the number" had cut off Mariko who at first did not continue following the interjection "oh." Mariko could have disregarded it and continued talking after the researcher had finished her utterance. Instead, she repeated saying "number." She said it as if speaking to herself or, to put it differently, as if she did not care who uttered which word. She moved on to say "this should be" (turn 06). Mariko's action produced a configuration of interaction having a feature of cognitive and emotional collaboration. A similar situation occurred in the next turn. Mariko was attempting to articulate the issue, but the researcher's saying "here" cut her off (turn 07). SungWon pointed to the bottom of the light intensity meter, where some numbers were written right above the two sockets, that is, the required information to know how to read the complex structure of the scale – the scale had a one fan-shaped graduation but two series of numbers above and below it. How these are used depends on which socket connects the meter to the sensor. SungWon continued: "there is a range, you know, you can select" (turn 07). This time again, Mariko did not insist on adding to her previous talk; instead, she said "Um, select" (turn 08).

Throughout the interaction, the two articulated more clearly that the issue concerned "the numbers" on the scale and what to "select." In this situation, SungWon's actions were more like those of a person participating in collaborative thinking, but at the same time SungWon was a researcher, not a fellow student who conducted the experiment. She said, "you can select" and by repeating "select" Mariko had a turn at figuring things out for herself. She looked at the meter for quite a long time without saying anything (turn 08). After a while, SungWon broke the silence and began to talk about how the left socket would be related to the numbers on the scale (turns 09–11). That SungWon spoke first implies that she understood Mariko's inaction as an expression that Mariko had not come to a clear understanding and needed to know more. In other words, Mariko's pause was achieved as an expression of an unclear understanding *by* the SungWon's action, and SungWon, *in* doing so, made her concern available for Mariko to see. The concern was achieved again when Mariko said "um" (turn 10) in response to the SungWon's "So if you select this one" (turn 09). If we were to see in SungWon's action of ending a pause as an attempt to be responsible for the situation, we may find one of its material grounds in the fact that Mariko produced her utterance ("select") by assuming the researcher's utterance ("you can select").

Mariko's pause led the situation to the SungWon's action of elaborating the sense of her "you can select." Whatever SungWon intended *by* saying "you can select," *in* saying so she was giving Mariko room to maneuver rather than requesting a specific action. In response, Mariko paused, which in turn gave SungWon room to maneuver. In both cases, the two did not force specific actions onto one another and therefore left open with respect to what and how to do next. The uncertainties involved in the actions allowed SungWon and Mariko opportunities to produce relevant acts that respected the other, thereby materializing their mutual care.

We therefore propose, as our third assertion, to take into account the development of ethics from a communicative perspective.

> *Assertion 3 A communicative act involves greater ethical value with respect to one another when it provides more opportunities for assuming another's acts. This gives all parties room to maneuver on their own grounds and*

thereby constitutes a process of producing new forms of configuration of humanness.

In the episode Mariko said to the researcher by the questioning sentence "do you know" (turn 02). Given her education and teaching experience, SungWon was the more knowledgeable relative to the instrumentation; the episode shows that knowledgeability is a matter of communicative interaction in a concrete situation. We saw that Mariko's actions of pause (turn 08) and an utterance (turn 09) had allowed the researcher to articulate her knowledgeability for the other. The researcher's action could become a manifestation of her knowledgeability in so far as it was a relevant and acceptable action in communication. After Mariko said "Oh, okay, okay," SungWon began to talk about the right socket. In this situation, however, interactions between the two were quite different from those of the previous situation in which the researcher had talked about the left socket (turns 09–12). SungWon began by uttering "if you" while pointing the right socket; but she paused for 0.57 second before continuing, "select this one." Mariko's response was also different. As SungWon said "this one" (turn 13), Mariko repeated "this one" and her hand had already pointed to the lower series of numbers on the scale (Figure 8.2e). SungWon's pause became a resource for Mariko to articulate her own knowledgeability for the other. In Mariko's gesture, SungWon's action achieved a concern for the Other, who might have already realized what would be talked and therefore might not needed more explanation.

In summary, the episode did not involve a mechanical, determinate process of asking, giving, and receiving information as if the speakers were placing words into prefigured slots. Researcher and participant collectively articulated and figured out the relevant issues at hand, which nevertheless had to unfold relevantly to one another in that they were communicating on different grounds. We consider every act as a manifestation of the way the other's act is assumed and reproduced on a continuously shifting ground. Every act with respect to the other constitutes an ethical configuration. Therefore, the issue is not what is asked and answered but how it is asked and answered.

TOWARD ETHICALLY VALUABLE RESEARCH ON LEARNING

> Saying is communication, to be sure, but as a condition for all communication, as exposure. Communication is not reducible to the phenomenon of truth and the manifestation of truth conceived as a combination of psychological elements . . . The elements of this mosaic are already in place in the antecedent exposure of the me to another, the non-indifference to the Other, which is not a simple "intention to address a message." The ethical sense of such an exposure to Another . . . is, from now on, visible. (Levinas, 1978, p. 82)

Levinas teaches us a lot about ethics. He suggests that saying is not just communication: it is exposure. Communication is not reducible to truth or to the psychologically conceived intention of speaking. Saying, communication, means non-indifference to the other, non-indifference with respect to participative understanding. Saying means exposure to Another; exposure is the essence of ethics and responsibility.

We begin this chapter with the articulation of an intention of developing a practical ethics for research on learning that is reflexive of the lived experiences and events that come about in the course of doing research. In response to it, we first present a moment of research on learning that was perceived as ethically critical (from a researcher's perspective) and articulate an initial issue into a dialectical question through a series of reflexive discussion. The initial question of "whether the researcher had to assist the participant's experiment" changes into "how the researcher and the participant experienced the situation collectively and learned to act with respect to the other." That is, at the very heart of the present issue is what, with Bakhtin (1993), we term *participative (unindifferent) understanding*. This form of understanding is unindifferent, because it is conscious of its effect on the other and therefore of the ontological responsibility for the other that we can only accept or deny but not eliminate from being. In the next step, the issue changes to the description of the event itself, particularly at the level of microanalysis; finally, we elaborate three assertions concerning research ethics. Throughout the chapter, we articulate different ways of talking "ethics in research on learning" that range from an ethically critical moment that requires more than following rules to a moment that is not critical. We exemplify how those different praxeologies were not only embedded in a praxis at different levels of reflexivity but also constitute themselves as a reflexive praxis that concretely realize possibilities inherent in lived experiences, that is, dialectical reflexivity. At this point, we ask "what is our ethics in the dialectical development of our study?" We respond to the question in terms of a following discussion toward ethically valuable learning research.

Ethics in research on learning is a matter of sensuous praxis and at the same time a matter of reflexive understanding of this praxis. Doing research, becoming part of participants' lifeworlds, means that researchers are conditioned by and therefore should deal with problematic situations as the participants experience them. Researchers cannot be direct problem solvers in and of those situations; but it is also true that they are not independent of the problems because they are also subjects in the same world. Doing research for us is a process of collectively overcoming problematic situations together with the participants, and at the same time producing knowledge that goes beyond localities of immediate and subjective experience. Despite the uncertain relations between events and theories, researchers can talk about learning phenomena because their pathic bodies enable them to experience and reflect. It is in recognizing the intersubjectivity of researcher and participant, who in their joint ongoing and concerted engagement constitute the research activity that we come to enact an ethical ethics rather than a political ethics that pretends to be apolitical.

Research on learning involves many facets including the learner's tasks, the researcher's participation, talking about the participation, and looking back on the talk. But accounts do not constitute lived experiences of doing research; the thrownness of first-hand experience is an irreducible and non-reproducible phenomenal field (e.g., Müller, 1972). Despite the encountered complexities, investigators do research for they can create new configurations of praxis having different qualities than any one before. Emerging praxis of different qualities comes with a reconfiguration of human bodies accompanying the evolution of communication, which people may call "becoming a researcher." Here, becoming is not a movement from existing as one to existing as another, but is the creation of some other that affirms both of them and at the same time leads to negations ultimately as dialectical continuation.

In this chapter we explicate the dialectical processes involved as an ethical issue came about and disappeared. That is, it became possible to take another look at the phenomenon as human subjects concretely produced the sensuous activity with and through their living/lived bodies, which we conceptualize as the dialectic of praxis and praxeology. Our study constitutes an ethical process of figuring out an ethical issue in that it is open to the developmental possibilities that emerge in the process. In research on learning, a researcher's ethical answerability lies in the opening up of possibilities to act in an answerable manner not only with respect to another but also with respect to herself. It is the answerability mediated by uncertainty and the equity established in one's own relevant ground for action.

Ethics in research on learning in science and mathematics is a reflexive endeavor. It establishes a researcher-participant interaction beneficial to the development of a participant's learning activities within research activities. This chapter shows that ethics develops as researchers concretize ethical aspects of their lived experiences in research activities, which realize ethical possibilities in those experiences and constitutes itself as a new configuration of praxis of ethics. That is, the dialectics of praxis and praxeology constitute the heart of a reflexive development of ethics.

EPILOGUE

KNOWING IN THE FLESH

> "Kant the philosopher," therefore, would have nothing to say about the flesh, the flesh of the philosopher, its "union of soul/mind and body." Nancy names, in effect, in a sequence that we have to reconstitute, "a relation of the body and thought about which Kant the philosopher has, in truth, *nothing* to say." (Derrida, 2000, p. 51)
>
> Any community is in essence affective and, by the same token, instinctive – and this does not only concern the fundamental communities of the society, the couple, the family, but all communities in general whatever their interests and explicit motivations. (Henry, 1990, p. 175)

Kant and all the constructivists following him, really have nothing to say about the relation of the body and thought. In fact, it is not the body generally that matters but the sensuous body, which is the source of knowing besides also being a material body. The living body, which in the phenomenological literature tends to be called *the flesh*, is the condition for making sense precisely because of its senses that are essential to affect and affectability. This living, sensuous body, the flesh, has not received much attention, if any, in science and mathematics education (but see Roth, 2010a, 2011a). The purpose of this book is to bring to science and mathematics educators' awareness the central role of the body in scientific and mathematical sense making and the associated development of higher-order cognitive functions from the perspective of the sensuous, living and lived body.

The sensuous, living and lived body is essentially characterized by affect. But, as Henry points out in the second introductory quote, affect also is the essence of any form of community, not only those forms of the fundamental kind, but of any community. For educators, community of practice and community of learning have been important ways for conceptualizing learning in classrooms. That is, we are held to think the very possibility of community in terms of the subjective experience of our sensuous, living and lived bodies.

Previous studies of human cognition have revealed the inseparable relation of mind and culture, on the one hand, and of body and mind, on the other (Roth, 2009). Science and mathematics educators have yet to attend to this new way of talking about knowing and learning, which is different from disembodied, computational, or information-processing models. We begin this book with a proposal for a dialectical approach to the body's central role in the evolutionary process of sense making. Throughout, we conduct analyses that exhibit the important role that the sensuous, living/lived body plays in knowing and learning science and mathematics. We support our assertions with concrete case examples in two main settings: second-grade elementary mathematics and third-year university physics. We conclude that a theory of the sensuous living/lived body is successful to the extent that it offers a new discourse for science and mathematics education research. The living body constitutes the answer for the ontological *more* that pertains to the very origin of sense making. Following phenomenological philosophers, we take the living body as constituting the first condition from which language, communication, and thinking appear and develop.

Sense making is possible because the living body is capable of hosting the cultural possibilities of communication and thinking, and thereby making living connections to other forms of bodies (e.g., representations) – not because bodies deliver information by themselves or because the human mind processes information like a computer. Talking about the living body directly addresses the first and foremost condition of knowing the world: the possibility to be affected physically and emotionally. In this epilogue, we further develop the theory of the living body and respond to the core problem of learning science or mathematics: Sense making in learning science and mathematics is not a metaphysical process but the living, "once-occurrent event of Being" (Bakhtin, 1993, p. 2). This affects the very condition of knowing the world not only for me but also to the Other. Learning science and mathematics makes me experience myself differently. That is, sense making involves the lived changes of my condition. But my life also is the life of the Other. When we take the perspective of life, we immediately transcend the opposition between you and me, between "them" and "us." I access life. I am life, in and through my sensuous, living/lived body. I exist *in flesh and blood.* The flesh, which is already part of the world, is the very capacity to sublate the division of the inside and the outside and lets the invisible rise through the ground and into the visible. This framework leads us to a very different way of talking about sense making and learning science and mathematics that leave behind the rationalist paradigms that manifest themselves today in the information processing and (social, radical) constructivist approaches.

THE FLESH AND BEING-WITH

> Community is nothing other than the ensemble of *living* individuals. . . . The attempt to oppose community and the individual, one to the other, to establish between them a hierarchical relation simply is non-sense, it takes us to oppose the *essence of life* to that which necessarily is implied by it. (Henry, 1990, p. 163, emphasis added)

Community, whether it is one of practice or learning, does not exist apart of *living* individuals. The essence of life is realized in and as the community of living individuals. In the quote that opens this section, the philosopher asks us not merely to consider the community and the individual in the abstract but both as modalities of life. Once we think community and the individual, collective and individual knowing, and human interactions from the perspective of life, we cannot but acknowledge the sensuous, living and lived body that makes all of these modalities possible. It is in and through the sensuous, living/lived body that I come to be as every other

I. It is precisely subjectivity and passibility that is common to all individuals. Being *in flesh and blood* means being-with others.

In what ways does the discourse of the living body allow us to better theorize sense making in science or mathematics education? In the introductory chapter of this book, we summarize the advantages of incorporating the living body thematic to sense making, each of which responds to a different key issue of research: (a) the inseparable relation of knowing and praxis, (b) the mutually constitutive relation of individual and collective, and (c) the unity of cognition and affect. Using our case studies in the three preceding parts of this book, we exemplify the capacity of the living body to constitute the foundation of learning and knowing. Three concepts are salient: passivity, heterogeneity, and solidarity. First, the living body is characterized by an essential passivity in a world including other bodies – it is given its intention and the absolutely new knowledge that it could not have oriented to and intended to learn. Second, the living body hybridizes cultural possibilities across multiple modalities and thus is heterogeneous. Third, the living body realizes the same life that other living bodies realize: It exists on collective grounds and affects it by its every action.

Following dialectical philosophy, we consider being always already as "being-with." This is the condition for the concept of solidarity – being is in such a mode in which my being constitutes the condition for the other in the way by which the other exists for my being (Nancy, 2000). Being-with supposes that being co-appears always already *with* the Other. I am what I am through the Other; Self and Other presuppose one another and achieves themselves what they are only through the other (plural singularity). For example, the case studies from the geometry lessons (chapters 1, 4, and 5) show the role of the audience in children's concept development. The generalized other is constitutive of what and who I am. Yet, this Other is never reducible to me – I do not "construct" the Other. The audience listens to my talk and, in and through its response, is constitutive of *what* I am saying. Intelligibility requires *participative (unindifferent) understanding*. This is why a dialectically appreciated being-with is equivalent to ethics. Being co-appears with the Other and is thus ethically related to the Other whom I affected in and through my action (Levinas, 1978). Therefore, being-with denotes plurality. It never speaks itself other than through a first-person "we are." Simultaneously, being-with is singular because it denotes *difference in itself*, the inner contradiction that makes the change with respect to itself. All being is *singular plural being* (Nancy, 2000). Focusing on the sensuous, living/lived body directly leads us into ethics.

The body of the human being is not simply the material (Ger., *Körper* and Fr. *corps*) but the sensuous and sensible *flesh* (i.e., Ger. *Leib* and Fr. *chair*). In the phenomenological sense, the notion of flesh is precisely related to the sense of owning a material body differentiated from another (i.e., flesh as my own body) and thus perceiving a foreign subjectivity, *the Other*. The flesh is singular in my body, but, because it constitutes life, it reflects the life of the Other who, in her flesh, shares the same life. The flesh constitutes a dialectical unit. In the flesh, we are simultaneously the same and other: "when *my* right hand touches *my* left hand .

. . the 'touching subject' passes over to the rank of the touched, descends into the things" (Merleau-Ponty, 1964, p. 174, our emphasis). The act of touching attests that I am in and through the Other. The living body constitutes the passage to the Other.

The centrality of the living body in being-with is clear in emotional aspects of communication. It is by the non-verbal means of communication that we come to share affect with others and which thereby serve as social glue (e.g., Collins, 2004). Feelings of emotion presuppose consciousness that objectifies the bodily state other than itself. For example, empathy (Ger. *Einfühlung*) denotes human capacity for understanding the Other and it has the same origin as feelings of emotion. This is so because what we share in and as life precisely is the capacity to be affected: affect (Henry, 2003). My capacity for experiencing the Other and feeling another's emotions corresponds to the plural nature of my living body; at the same time, the singular nature of my living body – my pain is always mine and not the pain another feels – corresponds to the capacity for not only finding myself feeling another's emotions but also confronting my own. Therefore, empathizing with an Other highlights the plurality of my living body. Because this singularity presupposes plurality, empathy allows me to take a third-person perspective on my emotions and thereby provides me with the possibility to expand my horizon of consciousness. It is in and through the flesh that I share life with others; it is in and through the flesh that I am affected; and it is in and through the affect that being-affected expresses itself.

SCIENTIFIC AND MATHEMATICAL LITERACY IN THE FLESH

Many educators think about understanding scientific or mathematical concepts in terms of abstract and formal thinking that somehow differs from the mundane ways that characterize everyday life. Doing science or mathematics somehow is something more elevated than walking, washing dishes, or talking about the weather. Yet without such mundane ways of being, no science or mathematics could have emerged (Husserl, 1939). For example, a complicated set of mathematical equations and calculations filling up a chalkboard in a physics classroom is a stereotypical symbol standing for the abstractness of science and mathematics. These are thought to exist independent of the bodily forms of human praxis. But in this form, mathematics and science exist only as virtuality – subjective human praxis is required for the production of the objective nature of mathematics and science. This abstraction of scientific and mathematical objects from their origin within real praxis is used as a justification for reducing sense making to a matter of skillful use of specific forms of representation and for the claim that only those who have special talents can achieve the requisite formal thinking. The chapters in this book show the significance of the living body, the flesh that is common to the human being in the teaching and learning at very different levels of development: in elementary mathematics (geometry) classes and university physics (thermodynamics) courses. Our analyses exhibit how mathematics and science come to be in the concrete, corporeal ways of the sensuous, living/lived body.

After all this talk about the flesh, some science and mathematics educators may ask what this has to do with thinking and whether the central role of the living body in sense making means that they have to teach gestures, body movements, body positions, and orientations to their students. The recurrent theme throughout this book is that the living body means more than having sensory organs of receiving information and sending it out to somewhere (like sensing machines). In terms of "more," we point out that it is in the flesh that the synthesis of different sense experiences is immediately achieved. Because the flesh constitutes an "I can," we do not need the mind for bringing different forms of experience together and "construct" them as belonging to the same subject. Kant has attempted to do this, and, as a result, needed to theorize an excessively powerful mind. In our approach, the synthesis occurs immanently because of the integrality of the flesh (Henry, 2000). Our synthesis does not refer to the conscious reflection on the experience but to the organic unity within the living body as a whole. For example, Clara in chapter 1 changes her way of holding a sample sphere from with her fingertips enclosing around its circular boundary to with her palm covering part of its curved surface. The sequential combination of her action with the production of the sound [sfiə(r)] (i.e., the word "sphere") in the following episode exemplifies that this is not the change of her hands (skills) as a singular organ but the change in the organic body (the flesh) that keeps in touch with the world and feels at the forefront one's relevance to the life condition – this increases the possibility for the sound-word "sphere" to mark sense in the situation. The flesh is the locus where the experience of the world and cultural possibilities (e.g., geometry) encounter one another. Thus, coming to see what has not been visible before (e.g., the curved surface of a spherical object compared to the flat surface of a two-dimensional circle) is associated with this organic synthesis in the flesh. Vygotsky conceptualizes this as the encounter between two lines of development (language, thinking) in the human ontogenesis, to generate a new possibility for the living being (therefore also for the vision) rather than as a matter restricted to the function of a specific organ (e.g., manual skills).

The development of higher-order cognitive functions involves the encounter (organic synthesis) between the corporeal, incarnate engagement with material objects and the associated transformation of them at an ideal and ideational level. Children who do not yet know the optical principles of an overhead projector (i.e., projection) are indeed capable of developing geometrical thinking about the correspondence between a three-dimensional object and a two dimensional shadow (e.g., projection) as we show below by drawing on a case example that extends Episode 1.1 (the conversation about the hidden-from-view mystery object on the overhead projector the projection of which is visible as a shade on the wall). With this episode, we exemplify: (a) the student's development of thinking in communication is related to her corporeal engagement with the object and developing language-like (communicative) actions in the course and (b) how the teacher provides an opportunity to a corporeal synthesis by expanding the possibilities of experiencing the world of objects.

Episode 9.1

```
23 Clara:    [it's not a a sphere because]
             [((Clara is gazing a yellow sphere in her hands.
             She holds it in front of her and turns it around
             within her hands))]
24           [(2.7)]
             [((Clara stretches her left palm and puts it on the
             left side of the sphere. She quickly bends her fin-
             gers and covers the curved surface.))]
25           [this part]
             [((Clara *stretches her left palm again and **rubs
             it around the contact point with the sphere. She
             raises the sphere with her right hand and ***turns
             its outward surface to the outside))] *(Figure
             9.1a) **(Figure 9.1b) ***(Figure 9.1c)
26           [is it's kind of sticking out]
             [((Clara *curls her left hand and **brings it near
             to the outward surface.))] *(Figure 9.1d) **(Figure
```

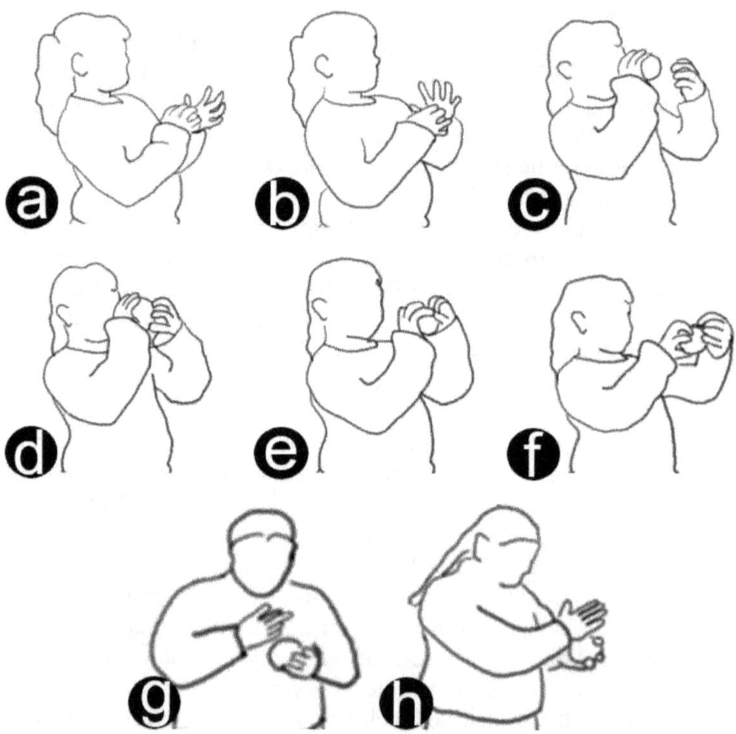

Figure 9.1. a. Clara stretches her left palm. b. She twists her plam around the contact point with the sphere. c. She raises the sphere with her right hand and turns its surface to the outside. d. She curves her left hand. e. She brings her curled hand near to the outward surface. f. She curls her left hand along the curved surface of the sphere. g. Daisy walks to the front curling her hand over the sphere. h. Daisy puts the edge of her palm on the sphere.

9.1e)
27 [and (that's it a reel?)]
[((*Clara *rolls her left hand along the curved surface*))] (Figure 9.1f)
28 Teacher: (°????°)
29 Teacher2: does everyone agree with that?
30 Student1: y:eah
31 Teacher: put your hand up if you agree with what Clara says
32 Student2: just a comment
33 Teacher: and those people who-
34 Teacher2: I, I, I think there was someone here with a different idea (????)
35 Teacher: Daisy?
36 Daisy: sphere↑
37 Teacher2: yeah
38 Teacher: and why do you think it's a sphere
39 Daisy: [°it's because: (??)°)]
[((*Daisy's gaze moves around the screen. Clara walks to Daisy and passes her the yellow sphere*))]
40 [(2.8)]
[((*Daisy receives the yellow sphere and stands up*))]
41 [because if you]
[((*Daisy walks to the front and turns toward the students. She stretched her right hand and *brings it over the sphere. She **places an edge of her palm on top of the sphere*))] *(Figure 9.1g) **(Figure 9.1h)
42 [cut up this way you'll have has um a circle (but?) (??)]
[((*Daisy turns her right hand around the curved surface and makes a circular movement*))]

Clara proposes the yellow sphere as a counterexample to the mystery object placed on the overhead projector by clarifying the difference between the sphere in her hands and the circle on the screen. Clara points out the curved surface of the sphere, which is different from the flatness of the circle on the screen despite the circular boundary common to both. In fact, a sphere may look like a circle from every angle if one focuses on the shape of the boundary. For those who do not know the sphere, even the shade usually added to a drawing of a sphere to make it look three-dimensional would not make a difference. In this situation, Clara interacts with the sphere from the initial act of touching and rubbing its curved surface with her palm and to the act of exposing the surface to the audience by raising it and juxtaposing her curved hand with it. This interaction between the sphere and her hands develops into the sign-like (deictic, iconic) action that she denotes by means of the expression of "sticking out." The "sticking out" of the surface of the sphere comes to be salient in Clara's vision. Clara may have the sense that the sphere is different from the circle when she raises her hand and grabs the yellow sphere from the chalkboard shelf (Episode 1.1). It is precisely here that she articulates what she feels and sees while holding the sphere and what constitutes the difference between the sphere and the circle. She experiences the three-dimensional nature of the sphere by letting herself be touched by the curved surface and the spatial volume between her palms. By touching the object with her hands, her whole body learns to experience the materiality that it is in contact with and her vision develops together with the touch – differentiation of the perfect round figures each in two- and three-dimensional space. Seeing what has not been salient and therefore invisible – e.g., curvedness – and the appearance of the object in consciousness are closely intertwined with the transformation in my interactions with a (visible, tangible) object. This also constitutes the transformation of my flesh (Derrida, 2000). It is in touching that I am touched by the world: materially, as the object touches my skin, and experientially, as I am lastingly changed in and through this encounter.

Clara's geometrical abstraction from the two juxtaposed objects remains underdeveloped. The image on the screen is a projected, two-dimensional shadow of the mystery object and therefore can never occupy a three-dimensional volume like a sphere. Therefore, her clarification of the difference does not necessarily mean that the mystery object placed on the overhead projector cannot be a sphere. Clara may be successful in articulating a particular (geometrical) aspect of the sphere (the outward curvedness) in relation to her claim, which for listeners also constitutes the material for their corporeal expressions and the corresponding thought. However, the essential problem here is to articulate the possibility or impossibility of the correspondence between the two objects in view of the projection that is not visible to the children. This invisible correspondence (projection) has to be made salient. This does not mean that children have to stop talking and learn about the optical principle of an overhead projector or the theory of projection. We see that children can bring their everyday experience to the conversation and expand the condition of their vision in the next situation. The two teachers present in the situation provide Daisy with an opportunity to articulate her claim – i.e., the mystery object might be a sphere. Daisy receives the sample sphere from Clara and exhibits the possibility for seeing the circular cross section of the sphere. She vertically holds her right palm over the sphere and produces a circular hand trajectory of cutting along the curved surface. Her iconic action of cutting the sphere together with the utterance "cut up" makes a link between the everyday experience of seeing a cross-section in a sphere-like object (e.g., cutting a watermelon) on the one hand and the sphere-circle correspondence on the other hand. Still, Daisy's proposal is underdeveloped since the overhead projector projection does not show any cross section but the largest cross-section. However, Daisy adds "cutting a sphere in half and seeing its cross-section" to the set of action possibilities available in this classroom, and thereby expands the possibility to see the invisible geometrical projection between the two figures.

This analysis exemplifies how learning involves the double development of the visible and the invisible. Children come to see the invisible relation between the sphere and the circle as they develop their interaction with the visible objects (i.e., touching). That is, the flesh is central to the transformation of the visible and coming to see the invisible. Simultaneously solidarity is at the heart of this double development that arises from the capacity of the flesh to synthesize the experience of the world in the flesh and crossing it with cultural possibilities. Clara actively follows the materiality of the object (sphere) and learns how to feel the curvedness of the surface instead of staying with feeling the round boundary. She thereby concretizes the sense of selecting a sphere among many geometrical objects, which increases the relevance of her action to the classroom conversation. The development of her experience of the object is contingent, but it is not arbitrary. She learns how to feel the object to expand its relevance to this conversation. In the same way, Daisy contingently mobilizes a new possibility to act toward a sphere in response to the teacher's questioning. The relevance of her proposal comes about not by her assertion itself but by its contribution to putting flesh to the invisible visible. Teachers assist students in coming to see the invisible not by attempting to directly visualize the invisible but by giving students opportunities to actively touch the material (visual) world and transform it in their flesh.

We begin this book with our proposal to consider the sensuous, living and lived body as the seat and origin of everyday rationality. We articulate a framework that is inspired by two lines of thought: cultural-historical, dialectical materialist psychology in the lineage that Lev Vygotsky initially proposed and materialist phenomenology in the lineage of Merleau-Ponty and Michel Henry. We articulate two developmental lines of scientific and mathematical literacy: communication and the sensuous experience of the material world. The consequence of this approach is that the encounter of these two lines of development appears as the growth point of sense making and cultural development. Moreover, this encounter of two lines enables the ontogenetic development of an individual because both are integral to life, the essence of which is the flesh. A theory of the flesh and of incarnation opens new avenues for thinking literacy and to develop a culturally responsive education.

REFERENCES

Austin, J. L. (1962). *How to do things with words.* Cambridge: Harvard University Press.

Bakhtin, M. M. (1981). *The dialogic imagination: Four essays.* Austin: University of Texas Press.

Bakhtin, M. M. (1993). *Toward a philosophy of the act.* Austin: University of Texas Press.

Bakhtine, M. [Volochinov, V. N.] (1977). *Le marxisme et la philosophie du langage: essai d'application de la méthode sociologique en linguistique* [Marxism and the philosophy of language: Essay on the application of the sociological method in linguistics]. Paris: Les Éditions de Minuit.

Bhatt, S. (1988). *Brunizem.* Manchester, UK: Carcanet

Bourdieu, P. (1997). *Méditations pascaliennes* [Pascalian meditations]. Paris: Seuil.

Brentano, F. (1924). *Psychologie vom empirischen Standpunkt* [Psychology from an empirical standpoint] (2nd enl. ed. by O. Kraus). Leipzig: Meiner. (First published in 1874)`

Collins, R. (2004). *Interaction ritual chains.* Princeton, NJ: Princeton University Press.

Derrida, J. (1967). *L'écriture et la différence* [Writing and difference]. Paris: Seuil.

Derrida, J. (1995). *The gift of death* (Translated by D. Wills). Chicago: University of Chicago Press.

Derrida, J. (1996). *Le monolinguisme de l'autre ou la prothèse d'origine* [Monolingualism of the other; or the prosthesis of origin]. Paris: Galilée.

Derrida, J. (2000). *Sur le toucher – Jean-Luc Nancy* [On touching – Jean-Luc Nancy]. Paris: Galilée.

Devereux, G. (1967). *From anxiety to method in the behavioral sciences.* The Hague: Mouton.

Eco, U. (1976). *A theory of semiotics.* Bloomington: Indiana University Press.

Foucault, M. (1975). *Surveiller et punir* [Discipline and punish]. Paris: Gallimard.

Garfinkel, H., & Sacks, H. (1986). On formal structures of practical action. In H. Garfinkel (Ed.), *Ethnomethodological studies of work* (pp. 160–193). London: Routledge & Kegan Paul.

Guba, E. G., & Lincoln, Y. S. (1989). *Fourth generation evaluation.* Newbury Park, CA: Sage.

Hegel, G. W. F. (1977). *Phenomenology of spirit* (Translated by A. V. Miller). Oxford: Oxford University Press.

Heidegger, M. (1954). *Was heißt Denken?* [What is called thinking]. Tübingen: Max Niemeyer.

Heidegger, M. (1977). *Sein und Zeit* [Being and time]. Tübingen: Max Niemeyer. (First published in 1927)

Henry, M. (1975). *Philosophy and phenomenology of the body* (Translated by G. Etzkorn). La Hague: Martinus Nijhoff.

Henry, M. (1990). *Phénoménologie matérielle* [Material phenomenology]. Paris: Presses Universitaires de France.

Henry, M. (2000). *Incarnation: Une phénoménologie de la chair* [Incarnation: A phenomenology of the flesh]. Paris: Seuil.

Henry, M. (2003). *Phénoménologie de la vie, tome II: De la subjectivité* [Phenomenology of life vol. 2: On subjectivity]. Paris: Presses Universitaires de France.

Holzkamp, K. (1991). Experience of self and scientific objectivity. In C. W. Tolman & W. Maiers (Eds.), *Critical psychology: Contributions to an historical science of the subject* (pp. 65–80). Cambridge: Cambridge University Press.

Husserl, E. (1939). Die Frage nach dem Ursprung der Geometrie als intentional-historisches Problem [The question about the origin of geometry as intentional-historical problem]. *Revue internationale de philosophie, 1,* 203–225.

Husserl, E. (1940). Grundlegende Untersuchungen zum phänomenologischen Ursprung der Räumlichkeit der Natur [Foundational investigations to the phenomenological origin of the spatiality of nature]. In M. Farber (Ed.), *Philosophical essays in memory of Edmund Husserl* (pp. 307–325). Cambridge, MA: Harvard University Press.

Husserl, E. (1980). *Vorlesungen zur Phänomenologie des inneren Zeitbewußtseins* [Lectures on the phenomenology of internal time consciousness]. Tübingen: Max Niemeyer. (First published in 1928)

Janvier, C. (1987). Translation processes in mathematics education. In C. Janvier (Ed.), *Problems of representation in the teaching and learning of mathematics* (pp. 27–32). Hillsdale, NJ: Lawrence Erlbaum Associates.

Kress, G., Jewitt, C., Ogborn, J., & Tsatsarelis, C. (2001). *Multimodal teaching and learning: The rhetorics of the science classroom*. London: Continuum.

Lakoff, G., & Núñez, R. E. (2000). *Where mathematics comes from: How the embodied mind brings mathematics into being*. New York: Basic Books.

Lave, J. (1997). On learning. *Forum Kritische Psychologie, 38*, 120–135.

Leont'ev, A. N. (1978). *Activity, consciousness and personality*. Englewood Cliffs, NJ: Prentice Hall.

Levinas, E. (1971). *Totalité et infini: Essai sur l'extériorité* (4ième ed.) [Totality and infinity: Essay on exteriority]. La Haye: Martinus Nijhoff.

Levinas, E. (1978). *Autrement qu'être ou au-delà de l'essence* [Otherwise than being or beyond essence]. La Haye: Martinus Nijhoff.

Levy, E., Duncan, S., & Cassell, J. (2007). Introduction: The dynamic dimension of language. In S. D. Duncan, J. Cassell, & E. T. Levy (Eds.), *Gesture and the dynamic dimension of language: Essays in honor of David McNeill* (pp. 3–11). Amsterdam: John Benjamins.

Marion, J.-L. (2004). *The crossing of the visible*. Stanford, CA: Stanford University Press.

Marx, K./Engels, F. (1962). *Werke Band 23: Das Kapital* [Works vol. 23: Capital]. Berlin: Dietz.

McNeill, D. (2002). Gesture and language dialectic. *Acta Linguistica Hafniensia, 34*, 7–37.

McNeill, D. (2005). *Gesture and thought*. Chicago: University of Chicago Press.

Merleau-Ponty, M. (1945). *Phénoménologie de la perception* [Phenomenology of perception]. Paris: Gallimard.

Merleau-Ponty, M. (1964). *Le visible et l'invisible* [The visible and the invisible]. Paris: Gallimard.

Merleau-Ponty, M. (1969). *La prose du monde* [The prose of the world]. Paris: Gallimard. (English translation first published in 1973)

Merleau-Ponty, M. (1996). *Sens et non-sens* [Sense and nonsense]. Paris: Gallimard. (First published in 1966)

Merleau-Ponty, M. (2001). *The incarnate subject: Malebranche, Biran, and Bergson in the union of the body and soul*. Amherst, NY: Humanity Books.

Meshcheryakov, A. (1979). *Awakening to life*. Moscow: Progress.

Moros, D. (2005). Production of scientific knowledge and radical uncertainty: The limits of the normative approach in innovation economics. *European Journal of Law and Economics, 20*, 305–322.

Müller, A. M. K. (1972). *Die präparierte Zeit: Der Mensch in der Krise seiner eigenen Zielsetzung* [Prepared time: Man in the crisis of his own goals]. Stuttgart, Germany: Radius Verlag.

Nancy, J.-L. (1993a). L'éloge de la mêlée [Eulogy of the mêlée]. *Transeuropéenne, 1*, 8–18.

Nancy, J.-L. (1993b). *The birth to presence* (Translated by B. Holmes and others). Stanford: CA, Stanford University Press.

Nancy, J.-L. (2000). *Being singular plural* (Translated by R. D. Richardson & A. E. O'Byrne). Stanford: CA, Stanford University Press.

Nancy, J.-L. (2008). *The discourse of the syncope: Logodaedalus*. Stanford, CA: Stanford University Press.

Ochs, E., Gonzales, P., & Jacoby S. (1996). "When I come down I'm in the domain state": Grammar and graphic representation in the interpretive activity of physicists. In E. Ochs, E. Schegloff, & S. Thompson (Eds.), *Interaction and grammar* (pp. 328–369). Cambridge: Cambridge University Press.

OED (2007). *Oxford English Dictionary*. http://dictionary.oed.com/

Plato (380 BCE). *Meno* (Translated by B. Jowett). The Internet Classic Archive. Retrieved February 24, 2010 from http://classics.mit.edu//Plato/meno.html.

Pozzer-Ardenghi, L., & Roth, W.-M. (2010). *Staging and performing scientific concepts*. Rotterdam: Sense Publishers.

Ricœur, P. (1991). *From text to action*. Evanston, IL: Northwestern University Press.

Ricœur, P. (1992). *Oneself as another* (Translated by K. Blamey). Chicago: University of Chicago Press. (First published in 1990)

Ricœur, P. (2004). *Sur la traduction* [On translation]. Paris : Bayard.

Rizzolatti, G., Fogassi, L., & Gallese, V. (2006). Mirrors in the mind. *Scientific American, 295*(5), 54–61.

Rorty, R. (1989). *Contingency, irony, and solidarity*. Cambridge: Cambridge University Press.

Roth, W.-M. (2002). *Being and becoming in the classroom*. Westport, CT: Ablex.

Roth, W.-M. (2008). Bricolage, métissage, hybridity, heterogeneity, diaspora: Concepts for thinking science education in the 21st century. *Cultural Studies of Science Education, 3*, 891–916.

Roth, W.-M. (2009). Appreciating the embodied social nature of mathematical cognition. In W.-M. Roth (Ed.), *Mathematical representation at the interface of body and culture* (pp. 335–350). Charlotte, NC: Information Age Publishing.

Roth, W.-M. (2010a). Incarnation: Radicalizing the embodiment of mathematics. *For the Learning of Mathematics, 30*(2), 8–18.

Roth, W.-M. (2010b). *Language, learning, context: Talking the talk*. London: Routledge.

Roth, W.-M. (2011a). *Geometry as objective science in elementary classrooms: Mathematics in the flesh*. New York: Routledge.

Roth, W.-M. (2011b). *Passibility: At the limits of the constructivist metaphor*. Dordrecht, The Netherlands: Springer.

Roth, W.-M. (2011c). Researching living/lived mathematical work. *Forum Qualitative Sozialforschung/ Forum Qualitative Social Research, 11*(1). Available at http://www.qualitative-research.net/index.php/fqs/article/view/1604

Roth, W.-M. (in press). Rules of bending, bending rules: The geometry of conduit bending in college and workplace. *Educational Studies in Mathematics*.

Roth, W.-M., & Barton, A. C. (2004). *Rethinking scientific literacy*. New York: Routledge.

Roth, W.-M., & Lawless, D. (2002a). Signs, deixis, and the emergence of scientific explanations. *Semiotica, 138*, 95–130.

Roth W.-M., & Lawless, D. (2002b). When up is down and down is up: Body orientation, proximity and gestures as resources for listeners. *Language in Society, 31*, 1–28.

Roth, W.-M., Lawless, D., & Tobin, K. (2000). Time to teach: Towards a praxeology of teaching. *Canadian Journal of Education, 25*, 1–15.

Roth, W.-M., McRobbie, C., Lucas, K. B., & Boutonné, S. (1997). The local production of order in traditional science laboratories: A phenomenological analysis. *Learning and Instruction, 7*, 107–136.

Roth, W.-M., & Middleton, D. (2006). The making of asymmetries of knowing, identity, and accountability in the sequential organization of graph interpretation. *Cultural Studies of Science Education, 1*, 11–81.

Roth, W.-M., & Pozzer-Ardenghi, L. (2006). Tracking situated, distributed, and embodied communication in real time. In M. A. Vanchevsky (Ed.), *Focus on cognitive psychology research* (pp. 237–261). Hauppauge, NY: Nova Science.

Roth, W.-M., & Thom, J. (2009a). Bodily experience and mathematical conceptions: From classical views to a phenomenological reconceptualization. *Educational Studies in Mathematics, 70*, 175–189.

Roth, W.-M., & Thom, J. (2009b). The emergence of 3d geometry from children's (teacher-guided) classification tasks. *Journal of the Learning Sciences, 18*, 45–99.

Roth, W.-M., & Tobin, K. (1996). Aristotle and natural observation versus Galileo and scientific experiment: An analysis of lectures in physics for elementary teachers in terms of discourse and inscriptions. *Journal of Research in Science Teaching, 33*, 135–157.

Roth, W.-M., Tobin, K., Carambo, C., & Dalland, C. (2005). Coordination in coteaching: Producing alignment in real time. *Science Education, 89*, 675–702.

Roth, W.-M., Tobin, K., Zimmermann, A., Bryant, N., & Davis, C. (2002). Lessons on/from the dihybrid cross: An activity theoretical study of learning in coteaching. *Journal of Research in Science Teaching, 39*, 253–282.

Roth, W.-M., & van Eijck, M. (2010). Fullness of life as minimal unit: Science, technology, engineering, and mathematics (STEM) learning across the life span. *Science Education, 94*, 1027–1048.

Roth, W.-M., Woszczyna, C., & Smith, G. (1996). Affordances and constraints of computers in science education. *Journal of Research in Science Teaching, 33*, 995–1017.

Sartre, J.-P. (1943). *L'être et le néant: Essai d'ontologie phénoménologique* [Being and nothingness: Essay on phenomenological ontology]. Paris: Gallimard.

Schutz, A. (1996). *Collected papers vol. iv*. Dordrecht: Kluwer Academic Publishers.

Sheets-Johnstone, M. (2009). *The corporeal turn: An interdisciplinary reader*. Exeter, UK: Imprint Academic.

Smith, D. E. (1990). *The conceptual practices of power: A feminist sociology of knowledge*. Toronto: University of Toronto Press.

Smith, D. E. (1999). *Writing the social: Critique, theory and investigations*. Toronto: University of Toronto Press.

Vygotsky, L. S. (1986). *Thought and language*. Cambridge, MA: MIT Press. (First published in 1934)

Vygotsky, L. S. (1989). Concrete human psychology. *Soviet Psychology, 27*(2), 53–77.

Waldenfels, B. (2004). Bodily experience between selfhood and otherness. *Phenomenology and the Cognitive Sciences, 3*, 235–248.

Waldenfels, B. (2006). *Grundmotive einer Phänomenologie des Fremden* [Fundamental motives of a phenomenology of the foreign/strange]. Frankfurt: Suhrkamp.

Wittgenstein, L. (1997). *Philosophische Untersuchungen/Philosophical investigations* (2nd ed.). Oxford: Blackwell. (First published in 1953)

INDEX

ABOUT THE AUTHORS

SungWon HWANG is assistant professor of science education at National Institute of Education, Nanyang Technological University, Singapore. She began her educational research career as middle school science teacher. She taught 7th, 8th, and 9th grade science for eight years in Korea, which allowed her spending a considerable amount of time designing scientific experiments and inquiries with her students. This therefore led her to investigate the role of the practical engagement in science laboratories in learning physics. She conducted her postdoctoral research at the University of Victoria, Canada. In the research laboratory of Professor Wolff-Michael Roth, she studied interdisciplinary theories and methods for better understanding the concrete dynamics of learning science and mathematics. Her recent publications focus on studying Vygotskian perspectives of language and identity and the role of the body in the development of literacy.

Wolff-Michael ROTH is Lansdowne Professor of Applied Cognitive Science at the University of Victoria, Canada. He conducts studies of learning across the entire lifespan, from multiple disciplinary perspectives and in different conceptual domains. He has published over 320 articles in peer-reviewed publications, more than 140 chapters in books, and 42 books published or in press. His special interests include empirical studies and theories that do not reduce knowing to a metaphysical mind or to social determinism. Thus, his recent work attempts to show how the dichotomies can be overcome. He is the author of *Dialogism: A Bakhtinian Perspective on Science Language and Learning* (Sense Publishers, 2009) and *Language, Learning, Context: Talking the Talk* (Routledge, 2010). He edited *Science Education from People for People: Taking a (Stand)point* (Routledge, 2009), *Mathematical Representation at the Interface of Body and Culture* (Information Age Publishers, 2009), and *Re/thinking Science Education: ReUniting Sociological and Psychological Perspectives* (Springer-Verlag, 2010).

www.ingramcontent.com/pod-product-compliance
Lightning Source LLC
Chambersburg PA
CBHW071448230926
51846CB00001B/383
* 9 7 8 9 4 6 0 9 1 5 6 5 9 *